John Henry Walsh

Mathematics for Common Schools

A Manual for Teachers, including Definitions, Principles, and Rules and

Solutions of the More Difficult Problems

John Henry Walsh

Mathematics for Common Schools
A Manual for Teachers, including Definitions, Principles, and Rules and Solutions of the More Difficult Problems

ISBN/EAN: 9783337157661

Printed in Europe, USA, Canada, Australia, Japan

Cover: Foto ©Paul-Georg Meister /pixelio.de

More available books at **www.hansebooks.com**

MATHEMATICS FOR COMMON SCHOOLS

A
MANUAL FOR TEACHERS

INCLUDING

DEFINITIONS, PRINCIPLES, AND RULES
AND SOLUTIONS OF THE MORE
DIFFICULT PROBLEMS

BY

JOHN H. WALSH

ASSOCIATE SUPERINTENDENT OF PUBLIC INSTRUCTION
BROOKLYN, N.Y.

INTERMEDIATE ARITHMETIC

BOSTON, U.S.A.
D. C. HEATH & CO., PUBLISHERS
1896

CONTENTS

(INTERMEDIATE ARITHMETIC MANUAL.)

I

                                                                          PAGE
INTRODUCTORY 1
    Plan and scope of the work — Grammar school algebra — Constructive geometry.

II

GENERAL HINTS 5
    Division of the work — Additions and omissions — Oral and written work — Use of books — Conduct of the recitations — Drills and sight work — Definitions, principles, and rules — Language — Analysis — Objective illustrations — Approximate answers — Indicating operations — Paper vs. slates.

IX

NOTES ON CHAPTER SIX 45

X

NOTES ON CHAPTER SEVEN 58

XI

NOTES ON CHAPTER EIGHT 68

XII

NOTES ON CHAPTER NINE 75

XIII

NOTES ON CHAPTER TEN 87

DEFINITIONS, PRINCIPLES, AND RULES i

ANSWERS 1

MANUAL FOR TEACHERS

I

INTRODUCTORY

Plan and Scope of the Work. — In addition to the subjects generally included in the ordinary text-books in arithmetic, *Mathematics for Common Schools* contains such simple work in algebraic equations and constructive geometry as can be studied to advantage by pupils of the elementary schools.

The arithmetical portion is divided into thirteen chapters, each of which, except the first, contains the work of a term of five months. The following extracts from the table of contents will show the arrangement of topics:

First and Second Years

Chapter I. — Numbers of Three Figures. Addition and Subtraction.

Third Year

Chapters II. and III. — Numbers of Five Figures. Multipliers and Divisors of One Figure. Addition and Subtraction of Halves, of Fourths, of Thirds. Multiplication by Mixed Numbers. Pint, Quart, and Gallon; Ounce and Pound. Roman Notation.

Fourth Year

Chapters IV. and V. — Numbers of Six Figures. Multipliers and Divisors of Two or More Figures. Addition and Subtraction of Easy Fractions. Multiplication by Mixed Numbers. Simple Denominate Numbers. Roman Notation.

Fifth Year

Chapters VI. and VII. — Fractions. Decimals of Three Places. Bills. Denominate Numbers. Simple Measurements.

Sixth Year

Chapters VIII. and IX. — Decimals. Bills. Denominate Numbers. Surfaces and Volumes. Percentage and Interest.

Seventh Year

Chapters XI. and XII. — Percentage and Interest. Commercial and Bank Discount. Cause and Effect. Partnership. Bonds and Stocks. Exchange. Longitude and Time. Surfaces and Volumes.

Eighth Year

Chapters XIII. and XIV. — Partial Payments. Equation of Payments. Annual Interest. Metric System. Evolution and Involution. Surfaces and Volumes.

While all of the above topics are generally included in an eight years' course, it may be considered advisable to omit some of them, and to take up, instead, during the seventh and eighth years, the constructive geometry work of Chapter XVI. Among the topics that may be dropped without injury to the pupil are Bonds and Stocks, Exchange, Partial Payments, and Equation of Payments.

Grammar School Algebra. — Chapter X., consisting of a dozen pages, is devoted to the subject of easy equations of one unknown quantity, as a preliminary to the employment of the equation in so much of the subsequent work in arithmetic as is rendered more simple by this mode of treatment. To teachers desirous of dispensing with rules, sample solutions of type examples, etc., the algebraic method of solving the so-called "problems" in percentage, interest, discount, etc., is strongly recommended.

In Chapter XV., intended chiefly for schools having a nine years' course, the algebraic work is extended to cover simple equations containing two or more unknown quantities, and pure and affected quadratic equations of one unknown quantity.

No attempt has been made in these two chapters to treat algebra as a science; the aim has been to make grammar-school pupils acquainted, to some slight extent, with the great instrument of mathematical investigation, — the equation.

Constructive Geometry. — Progressive teachers will appreciate the importance of supplementing the concrete geometrical instruction now given in the drawing and mensuration work. Chapter XVI. contains a series of problems in construction so arranged as to enable pupils to obtain for themselves a working knowledge of all the most important facts of geometry. Applications of the facts thus ascertained, are made to the mensuration of surfaces and volumes, the calculation of heights and distances, etc. No attempt is made to anticipate the work of the high-school by teaching geometry as a science.

While the construction problems are brought together into a single chapter at the end of the book, it is not intended that instruction in geometry should be delayed until the preceding work is completed. Chapter XVI. should be commenced not later than the seventh year, and should be continued throughout the remainder of the grammar-school course. For the earlier years, suitable exercises in the mensuration of the surfaces of triangles and quadrilaterals, and of the volumes of right parallelopipedons have been incorporated with the arithmetic work.

II

GENERAL HINTS

Division of the Work. — The five chapters constituting Part I. of *Mathematics for Common Schools* should be completed by the end of the fourth school year. Chapter I., with the additional oral work needed in the case of young pupils, will occupy about two years; the remaining four chapters should not take more than half a year each. When the Grube system is used, and the work of the first two years is exclusively oral, it will be possible, by omitting much of the easier portions of the first two chapters, to cover, during the third year, the ground contained in Chapters I., II., and III. The remaining eight arithmetic chapters constitute half-yearly divisions for the second four years of school.

Additions and Omissions. — The teacher should freely supplement the work of the text-book when she finds it necessary to do so; and she should not hesitate to leave a topic that her pupils fully understand, even though they may not have worked all the examples given in connection therewith. A very large number of exercises is necessary for such pupils as can devote a half-year to the study of the matter furnished in each chapter. In the case of pupils of greater maturity, it will be possible to make more rapid progress by passing to the next topic as soon as the previous work is fairly well understood.

Oral and Written Work. — The heading "Slate Problems" is merely a general direction, and it should be disregarded by the teacher when the pupils are able to do the work "mentally." The use of the pencil should be demanded only so far as it may be required. It is a pedagogical mistake to insist that all of the pupils of a class should set down a number of figures that are not needed by the brighter ones. As an occasional exercise, it may be advisable to have scholars give all the work required to solve a problem, and to make a written explanation of each step in the solution; but it should be the teacher's aim to have the majority of the examples done with as great rapidity as is consistent with absolute correctness. It will be found that, as a rule, the quickest workers are the most accurate.

Many of the slate problems can be treated by some classes as "sight" examples, each pupil reading the question for himself from the book, and writing the answer at a given signal without putting down any of the work.

Use of Books. — It is generally recommended that books be placed in pupils' hands as early as the third school year. Since many children are unable at this stage to read with sufficient intelligence to understand the terms of a problem, this work should be done under the teacher's direction, the latter reading the questions while the pupils follow from their books. In later years, the problems should be solved by the pupils from the books with practically no assistance whatever from the teacher.

Conduct of the Recitation. — Many thoughtful educators consider it advisable to divide an arithmetic class into two sections, for some purposes, even where its members are nearly equal in attainments. The members of one division of such a class may work examples from their books while the others write the answers to oral problems given by the teacher, etc.

Where a class is thus taught in two divisions, the members of each should sit in alternate rows, extending from the front of the room to the rear. Seated in this way, a pupil is doing a different kind of work from those on the right and the left, and he would not have the temptation of a neighbor's slate to lead him to compare answers.

As an economy of time, explanations of new subjects might be given to the whole class; but much of the arithmetic work should be done in "sections," one of which is under the immediate direction of the teacher, the other being employed in "seat" work. In the case of pupils of the more advanced classes, "seat" work should consist largely of "problems" solved without assistance. Especial pains have been taken to so grade the problems as to have none beyond the capacity of the average pupil that is willing to try to understand its terms. It is not necessary that all the members of a division should work the same problems at a given time, nor the same number of problems, nor that a new topic should be postponed until all of the previous problems have been solved.

Whenever it is possible, all of the members of the division working under the teacher's immediate direction should take part in all the work done. In mental arithmetic, for instance, while only a few may be called upon for explanations, all of the pupils should write the answers to each question. The same is true of much of the sight work, the approximations, some of the special drills, etc.

Drills and Sight Work.— To secure reasonable rapidity, it is necessary to have regular systematic drills. They should be employed daily, if possible, in the earlier years, but should never last longer than five or ten minutes. Various kinds are suggested, such as sight addition drills, in Arts. 3, 11, 24, 26, etc.; subtraction, in Arts. 19, 50, 53, etc.; multiplication, in Arts. 71, 109, etc.; division, in Arts. 199, 202, etc.; counting by 2's, 3's, etc., in Art. 61; carrying, in Art. 53, etc. For the young pupil, those are the most valuable in which the figures are in his sight, and in the position they occupy in an example; see Arts. 3, 34, 164, etc.

Many teachers prepare cards, each of which contains one of the combinations taught in their respective grades. Showing one of these cards, the teacher requires an immediate answer from a pupil. If his reply is correct, a new card is shown to the next pupil, and so on. Other teachers write a number of combinations on the blackboard, and point to them at random, requiring prompt answers. When drills remain on the board for any considerable time, some children learn to know the results of a combination by its location on the board, so that frequent changes in the arrangement of the drills are, therefore, advisable. The drills in Arts. 111, 112, and 115 furnish a great deal of work with the occasional change of a single figure.

For the higher classes, each chapter contains appropriate drills, which are subsequently used in oral problems. It happens only too frequently that as children go forward in school they lose much of the readiness in oral and written work they possessed in the lower grades, owing to the neglect of their teachers to continue to require quick, accurate review work in the operations previously taught. These special drills follow the plan of the combinations of the earlier chapters, but gradually grow more difficult. They should first be used as sight exercises, either from the books or from the blackboard.

To secure valuable results from drill exercises, the utmost possible promptness in answers should be insisted upon.

Definitions, Principles, and Rules. — Young children should not memorize rules or definitions. They should learn to add by adding, after being first shown by the teacher how to perform the operation. Those not previously taught by the Grube method should be given no reason for "carrying." In teaching such children to write numbers of two or three figures, there is nothing gained by discussing the local value of the digits. During the earlier years, instruction in the art of arithmetic should be given with the least possible amount of science. While principles may be incidentally brought to the view of the children at times, there should be no cross-examination thereon. It may be shown, for instance, that subtraction is the reverse of addition, and that multiplication is a short method of combining equal numbers, etc.; but care should be ta~~~ ~se of pupils below about the fifth school year not to dwell long on this side of the instruction. By that time, pupils should be able to add, subtract, multiply, and divide whole numbers; to add and subtract simple mixed numbers, and to use a mixed number as a multiplier or a multiplicand; to solve easy problems, with small numbers, involving the foregoing operations and others containing the more commonly used denominate units. Whether or not they can explain the principles underlying the operations is of next to no importance, if they can do the work with reasonable accuracy and rapidity.

When decimal fractions are taken up, the principles of Arabic notation should be developed; and about the same time, or somewhat later, the principles upon which are founded the operations in the fundamental processes, can be briefly discussed.

Definitions should in all cases be made by the pupils, their mistakes being brought out by the teacher through appropriate questions, criticisms, etc. Systematic work under this head should be deferred until at least the seventh year.

The use of unnecessary rules in the higher grades is to be deprecated. When, for instance, a pupil understands that *per cent* means *hundredths*, that seven per cent means seven hundredths, it should not be necessary to tell him that 7 per cent of 143 is obtained by multiplying 143 by .07. It should be a fair assumption that his previous work in the multiplication of common and of decimal fractions has enabled him to see that 7 per cent of 143 is $\frac{7}{100}$ of 143 or 143 × .07, without information other than the meaning of the term " per cent."

When a pupil is able to calculate that 15% of 120 is 18, he should be allowed to try to work out for himself, without a rule, the solution of this problem: 18 is what per cent of 120? or of this: 18 is 15% of what number? These questions should present no more difficulty in the seventh year than the following examples in the fifth: (*a*) Find the cost of $\frac{1}{20}$ ton of hay at $12 per ton. (*b*) When hay is worth $12 per ton, what part of a ton can be bought for $1.80? (c) If $\frac{3}{20}$ ton of hay costs $1.80, what is the value of a ton?

When, however, it becomes necessary to assist pupils in the solution of problems of this class, it is more profitable to furnish them with a general method by the use of the equation, than with any special plan suited only to the type under immediate discussion.

In the supplement to the Manual will be found the usual definitions, principles, and rules, for the teacher to use in such a way as her experience shows to be best for her pupils. The rules given are based somewhat on the older methods, rather than on those recommended by the author. He would prefer to omit entirely those relating to percentage, interest, and the like as being unnecessary, but that they are called for by many successful teachers, who prefer to continue the use of methods which they have found to produce satisfactory results.

Language. — While the use of correct language should be insisted upon in all lessons, children should not be required in arithmetic to give all answers in "complete sentences." Especially in the drills, it is important that the results be expressed in the fewest possible words.

Analyses. — Sparing use of analyses is recommended for beginners. If a pupil solves a problem correctly, the natural inference should be that his method is correct, even if he be unable to state it in words. When a pupil gives the analysis of a problem, he should be permitted to express himself in his own way. Set forms should not be used under any circumstances.

Objective Illustrations. — The chief reason for the use of objects in the study of arithmetic is to enable pupils to work without them. While counters, weights and measures, diagrams, or the like are necessary at the beginning of some topics, it is important to discontinue their use as soon as the scholar is able to proceed without their aid.

Approximate Answers. — An important drill is furnished in the "approximations." (See Arts. 521, 669, 719, etc.) Pupils should be required in much of their written work to estimate the result before beginning to solve a problem with the pencil. Besides preventing an absurd answer, this practice will also have the effect of causing a pupil to see what processes are necessary. In too many instances, work is commenced upon a problem before the conditions are grasped by the youthful scholar; which will be less likely to occur in the case of one who has carefully "estimated" the answer. The pupil will frequently find, also, that he can obtain the correct result without using his pencil at all.

Indicating Operations. — It is a good practice to require pupils to indicate by signs all of the processes necessary to the solution of a problem, before performing any of the operations. This frequently enables a scholar to shorten his work by cancellation, etc. In the case of problems whose solution requires tedious processes, some teachers do not require their pupils to do more than to indicate the operations. It is to be feared that much of the lack of facility in adding, multiplying, etc., found in the pupils of the higher classes is due to this desire to make work pleasant. Instead of becoming more expert in the fundamental operations, scholars in their eighth year frequently add, subtract, multiply, and divide more slowly and less accurately than in their fourth year of school.

Paper *vs*. Slates. — To the use of slates may be traced very much of the poor work now done in arithmetic. A child that finds the sum of two or more numbers by drawing on his slate the number of strokes represented by each, and then counting the total, will have to adopt some other method if his work is done on material that does not permit the easy obliteration of the tell-tale marks. When the teacher has an opportunity to see the number of attempts made by some of her pupils to obtain the correct quotient figures in a long division example, she may realize the importance of such drills as will enable them to arrive more readily at the correct result.

The unnecessary work now done by many pupils will be very much lessened if they find themselves compelled to dispense with the "rubbing out" they have an opportunity to indulge in when slates are employed. The additional expense caused by the introduction of paper will almost inevitably lead to better results in arithmetic. The arrangement of the work will be looked after; pupils will not be required, nor will they be permitted, to waste material in writing out the operations that can be performed mentally; the least common denominator will be determined by inspection; problems will be shortened by the greater use of cancellation, etc., etc. Better writing of figures and neater arrangement of problems will be likely to accompany the use of material that will be kept by the teacher for the inspection of the school authorities. The endless writing of tables and the long, tedious examples now given to keep troublesome pupils from bothering a teacher that wishes to write up her records, will, to some extent, be discontinued when slates are no longer used.

IX

NOTES ON CHAPTER SIX

The previous work in mixed numbers should make the pupils reasonably familiar with the addition and subtraction of fractions having small denominators. In this chapter, the work is extended to cover the addition and subtraction of fractions whose common denominator is determinable by inspection. For the present, the teacher should be satisfied if her pupils acquire reasonable facility in performing the various operations, even if they are unable to formulate, in the language of experienced mathematicians, the reasons for the different steps. The children should be required to use correctly and intelligently such technical terms as are required by the work of the chapter; but they should not be compelled to memorize any definitions that convey to them no meaning. They should incidentally learn what is meant by numerator, denominator, common denominator, multiple, etc., by hearing the teacher employ these words from time to time, rather than by commencing with what is to them an unintelligible jumble of words.

451. While systematic work in fractions belongs properly to the next chapter, the teacher should not hesitate to call $1\frac{2}{3}$, $1\frac{5}{4}$, etc., "improper fractions," and to ask a pupil to state how they are changed to whole or to mixed numbers.

453. Do not, for the present, formulate the rule for changing a fraction to an equivalent one with higher terms.

458. The meaning of "lowest terms" is given in No. 6. Leave the rule for the next chapter. After a pupil has reasoned out in his own way that 18 hours is $\frac{3}{4}$ day, in No. 15, the teacher may explain that 18 hours can be written $\frac{18}{24}$ day, which is reducible to the answer given above.

463. Have pupils see that $\frac{1}{6}$ is larger than either $\frac{1}{7}$ or $\frac{1}{8}$, because 1 sixth is larger than a seventh or an eighth; and this for the reason that the fewer the number of equal divisions made in a unit, the larger is each portion. Do not require scholars to change these fractions to equivalent ones having a common denominator.

467. For finding the difference between two mixed numbers when the fraction in the subtrahend is greater than that in the minuend, the method given in the text-book is the one generally employed. The teacher should always consider herself at liberty to use any other way of performing this and other operations, but she should not willingly adopt any method that is more tedious. Children should not, for instance, be required to change mixed numbers to improper fractions, and then to reduce these to a common denominator in order to subtract one from the other.

469. Pupils should now be required to pay more and more attention to the arrangement of the problem work, without, however, being permitted to use unnecessary figures or to waste time. In some good schools, the full written analysis of a problem is occasionally used as an exercise in composition.

When the pupils find difficulty in determining the operations necessary to the solutions of problems, the latter should be used as "sight" work. The alterations in the figures needed to simplify a problem should now be made by a pupil, instead of by the teacher, as recommended in previous chapters. The scholar that reads No. 1, for instance, might change $5\frac{1}{4}$ and $4\frac{2}{3}$ yards, to 5 and 4, respectively. No. 2 can be solved as it stands. In No. 3, $150 might be substituted for $140.40, and $2 for $1.80.

NOTES ON CHAPTER SIX 47

Work of this kind should gradually lead the pupil to form the habit of using some similar method of ascertaining for himself how to manage a problem.

471. Written in this form: $3\overline{)93\frac{2}{3}}$, No. 11 should give the children no trouble. If, however, they hesitate when the fraction is reached, the difficulty may be cleared up by making a concrete problem: Divide $93\frac{2}{3}$ equally among 3 persons. What is the share of each?

Under no circumstances should these dividends be changed to improper fractions.

484. In nearly all of the previous multiplication work involving mixed numbers, the latter have been used as multipliers. In No. 35, the mixed number appears as a multiplicand. The first six of these examples and the last two should be used as sight work, the answers being written directly from the book. When the pupil reaches one that needs to be worked out in full, say No. 41, he should not be permitted to use $18\frac{2}{3}$ as a multiplier, as it is important that he should learn the proper method of working both classes of examples.

$$\begin{array}{r} 18\frac{2}{3} \\ 13 \\ \hline 9\overline{)91} \\ 10\frac{1}{9} \\ 54 \\ 18 \\ \hline \end{array}$$

489. Many scholars will carelessly give 20 halves as the result obtained by dividing 500 halves by 25 halves. To prevent the possibility of a mistake of this kind, some teachers multiply the divisor and the dividend by the least common multiple of the denominators of the fractions. While this method produces exactly the same figures as the one given in the text-book, it is probably less likely to be followed by the error mentioned above.

$$\begin{array}{r} 18\frac{3}{4}\overline{)1387\frac{1}{2}} \\ \times 4 \quad \times 4 \\ \hline 75\overline{)5550} \end{array}$$

497. Some teachers may prefer to write the example as is here given, although using 5 as the multiplier. Other teachers "analyze" as follows: At 1¢ per lb., 157 pounds of sugar would cost $1.57; at 5¢ per lb., the

$$\begin{array}{r} \$.05 \\ \times 157 \\ \hline \$7.85 \end{array}$$

$1.57 cost is 5 times $1.57. Business men pay no attention
× 5   to these fine-spun distinctions; they use as a multiplier
───── the most convenient number, and write the dollar sign
$7.85
and the period in the product alone.

500. In analyzing problems of this kind, it is better, perhaps, to emphasize the fact that multiplication is employed in obtaining the result. Thus, 32 base-balls @ 25¢ = 32 times $¼ = 32 quarters = $8

502. In No. 28, the price of 11 yards can be found by taking 11 times $¾, or 33 quarters, etc. No. 24 is rendered easier by saying that 24 bushels at 1 quarter per bushel would cost 6 dollars; and that at 3 quarters per bushel the cost would be 3 times 6 dollars, etc. Pupils should be encouraged to use the method best adapted to the particular example under consideration.

509. See Art. 306.

510. In finding, for example, the number of 50-cent knives that can be purchased for $20, it may be advisable to make the division idea prominent. The analysis can take some such form as this: There can be bought as many knives as one half-dollar is contained times in 20 dollars — or, as there are half-dollars in 20 dollars.

Later problems involving division of fractions cause less trouble if the appropriate operation is always kept before the pupils, regardless of the method employed to shorten the solution of questions of certain types. These short methods should, however, be used.

To ascertain the number of 2-dollar knives obtainable for $24, the scholar turns naturally to division; and he should learn to see that he actually divides when he obtains 48 as the number of 50-cent knives that can be purchased for the same money. In the latter case, the numbers given are 24 and $\frac{1}{2}$, from which 48 can result only when $\frac{1}{2}$ is used as a divisor.

Many pupils that give the correct answer when $24 \div \frac{1}{2}$ is placed upon the blackboard as a sight example will think that 12 is the quotient of $\frac{1}{2}\overline{)24}$. For purposes of drill, this last form should occasionally be employed in sight work, as should be the third form of division, $\dfrac{24}{\frac{1}{2}}$.

511. Example 6: There can be bought as many bars of soap as there are quarters in $3¼, or 13 bars. Example 8: As many yards as there are quarters in $5¾. Example 9: As many bushels as there are quarters in $10¾.

While set forms of analysis should not be required in any grades, older pupils should be led to use such as are most likely to lead to an intelligent appreciation of mathematical principles. From the beginning of about the fifth year, the science of arithmetic should begin to receive some attention, but not so much as to lessen to too great an extent the time that should be devoted to arithmetic as an art.

516. These examples are introduced to lead up to division of Federal money. From their previous experience, the scholars will readily work the first example, for instance, by changing it to the form $24½ ÷ $½ = $\frac{49}{2} \div \frac{1}{2} = 49$. No. 2 becomes $12¼ ÷ $¼; No. 3, $26 ÷ $¼ = $\frac{78}{3} \div \frac{1}{3}$; etc.

Without laying much stress upon the terms "abstract" and "concrete," the teacher should bring her pupils to understand that the quotient of the first example is 49, — not 49 dollars.

517. While giving the answers to these exercises, the children should be able to state, after proper questioning, that the dividend must be of the same denomination as the divisor. In 2 ft. + 8 in., instead of changing the divisor to ⅔ ft., they naturally reduce the dividend to 24 inches, even if in 2 ft. + 6 in. they may have used ½ ft. as the divisor.

518. Changing the dividend to cents, No. 1 becomes 400 cents ÷ 10 cents, or 400 ÷ 10. To No. 11, many will give 50 as the result, unless previously well taught. In No. 12, the denomination of both terms being the same, the problem becomes $3 \div \frac{1}{4}$, or 12 quarters ÷ 1 quarter, rather than 300 ÷ 25. Nos. 13–20 are more readily worked by reducing each divisor to a fraction of a dollar. Pupils should understand that the answer is the same whether the dividend is changed to the same denomination as the divisor, or *vice versa*.

519. Some teachers write this example .36)27.00. It will be found safer to make the terms whole numbers by changing both to cents.

520. The method suggested for the first example, 11000 ÷ 275, is the more general one, although longer, perhaps, than $110 \div 2\frac{3}{4}$. Do not permit long division in No. 2. In No. 3, after writing 14000 ÷ 560, the pupils should strike out a cipher from each term: this should be insisted upon whenever the divisor ends in a cipher. In No. 4, either $74\frac{1}{2} \div \frac{1}{2}$ or $745\cancel{0} \div 5\cancel{0}$ should be accepted. If the work in No. 7 takes the form 75)$27.00, the answer should be $.36; if the pupil writes 75)2700¢, his answer should be 36 cents. The first form is the one employed in the work of preceding chapters, and no change should be suggested.

521. These drills in obtaining approximate results are intended to lead the pupil to such an examination of his answer as will prevent his being satisfied with one very much out of the way. It should not be expected that the same approximation will be obtained by all the members of a class.

2. 4200 ÷ 200.
3. $\frac{1}{2} \times 48$.
4. 12000 ÷ 2000.
5. $2 × 99, or $1.95 × 100, or $2 × 100.
6. 30 + 38.
7. 175 ÷ 25.
8. 19 × 10.
9. 87 − 50.
10. 5 × 5 × 5.

NOTES ON CHAPTER SIX

526. Formal instruction in denominate numbers should be deferred for a year or more. The average scholar will be able to solve all these problems if left to himself.

528-532. See notes on previous drills of this kind, Arts. 286 and 350. Special exercises in multiplication are regularly given by some teachers in the following manner:

2, 12, 22, 32, 42, 52, 62, 72, 82, 92
× 5

A horizontal or a vertical row of numbers ending in 2, for instance, is written on the board with, say, 5 as a multiplier. Attention is called to the fact that 2×5 is 10, so that all of these products must end in 0.

The pupils are also reminded that when the multiplicand is a number of two figures, 1 must be carried to the product of 5 times the tens' figure. When the teacher points to 12, the pupil says 5, 6, 60 — the first number (5) being the product of the multiplier and the tens' figure of the multiplicand; the second, (6) being this product increased by the carrying figure 1; the third being the result, which has been completed by annexing the units' figure (0) of the first product (2×5).

Pointing to 52, the pupil says 25, 26, 260; to 92, he says 45, 46, 460. After sufficient drill with 5 as a multiplier, it is replaced successively by 6, 7, 8, etc. The row of multiplicands is also changed to 3, 13, 23, etc.; 4, 14, 24, etc.; when the previous row has been employed with all of the multipliers, say from 5 to 12. With sufficient practice of this kind, pupils become able to give the product of any number of two figures, by multipliers to 12, with great readiness.

Some teachers, however, prefer in oral multiplication to use the method previously suggested of commencing to multiply at the tens' figure of the multiplicand. In finding 5 times 38, the pupil takes the latter number as it is given, thirty — eight, and multiplies in the same order, obtaining 150 and 40, or 190.

534–535. Long-division drills. See Arts. 321 and 397–401.

538. See Art. 563, p. 55, and Arithmetic, Art. 385.

540. See Art. 384.

543. In problems of this kind, writing the given numbers in the places called for by the conditions of each example helps the pupil in his solution. After writing No. 5, as here indi-

$$
\begin{array}{lll}
1. \;\; 68 & 2. \quad\;\; ? & 5. \quad\;\; ? \\
43 & 24\overline{)264} & -89 \\
\underline{\;?\;} & & \overline{\;\;92} \\
150 & &
\end{array}
$$

cated, he can see that addition is the required operation more readily than if he endeavors to determine it from the words of the problem.

546. Teachers should not weary pupils by giving too many items in the earlier bills. It is useful to employ occasionally such quantities and prices as will not require the use of a separate piece of paper to perform the necessary multiplications. In No. 1, for instance, the pupil should be compelled to fill out the cost of each item without recourse to his slate. If he does not know the product of 16×5, he should multiply one figure at a time, writing the result in its proper place. Except, possibly, No. 2, the other bills called for under this section should be made out in the way suggested for No. 1, the use of a slate or other paper not being permitted.

The form given in the text-book is the one generally followed by business men. The first two vertical lines are kept to enclose the day of the month (see Arithmetic, Art. 642). The total cost of each item is placed in the first columns of dollars and cents, the amount of the bill being placed in the last columns, and on the line below the last item. When a single article is sold, its cost is placed directly in the first columns of dollars and cents, and is not written in the column of prices. Unnecessary words, — *at* or @, for instance, *per yd.*, *lb.*, etc., — are never used; nor are commas employed after the names of the articles. It is now customary to omit the period after the date and after the name of the seller. The names of the articles are generally commenced with capitals, and the quantities are written with small letters. The heading given is the one most frequently used, though other forms are common; such as

ABRAHAM AND STRAUS

Sold to MRS. H. T. SHORT

Pupils should write the cost of 10¼ lb. of chicken @ 30¢, in No. 4, as $3.08. Fractions of cents should not appear in the results; those below ½¢ being rejected, and those of ½¢ and higher being considered 1¢.

547–551. When pupils have become familiar with the notation and numeration of decimals, the remaining decimal work of this chapter should not require much discussion.

554. After working Nos. 1–4, pupils should be left to themselves to arrange No. 5. Nearly all of them will place the numbers in their proper places. Neither require nor permit unnecessary ciphers to be employed to fill out all the numbers to three decimal places.

555. The above suggestions apply here.

559. In the product of .36 by 3, pupils will naturally place the decimal point where it belongs, as they will in example No. 42. Before working No. 43, they should be required to deduce the rule for "pointing off." All unnecessary ciphers should be canceled in the product, the answer to No. 52 being read by the pupil as 960, not 960 and no tenths — 960.0.

560. These exercises should lead scholars to see that the number of decimal places in the multiplicand cancels a corresponding number of ciphers in the multiplier.

561. In giving the quotient of No. 11, a pupil may write $9\frac{32}{100}$. If called upon to read this answer, he will see that 9.32 expresses the same result. He will readily understand that $\frac{88}{1000}$ is also written .086. After a few examples, he can state the rule.

From this point, the teacher may change the method of dividing by a number ending in one or more ciphers. Instead of marking off, by a line, a corresponding number of figures from the right of the dividend, the pupil can locate the decimal point in the proper place. In No. 17, the decimal point will be moved one place to the left, to divide by 10; moving it two places to the left in the dividend of No. 18 will give its quotient; etc.

563. The rule for "pointing off" should be deduced from the sight exercises of Art. 562. In dividing 8 and 64 hundredths by 2, the pupil will, without prompting, obtain 4 and 32 hundredths. When he comes to No. 5, 8.4 ÷ 5, he can be led to see that this example is the equivalent of No. 4, 8.40 ÷ 5, in which the quotient is 1.68. Nos. 9 and 10 also require the annexation of a cipher at the right of the dividend.

When the scholars understand that the quotient must contain the same number of decimal places as the dividend, including any ciphers that may have been annexed to the latter, they should be taught the method a business man would employ. The latter, in dividing 120 by 64, does not find it necessary to write ciphers in the dividend, and then to count the number thus annexed, in order to determine the position of the decimal point in the quotient. He sees at once that the result is 1 and a decimal, and he places the point after the 1, before he writes the next quotient figure.

In a short-division example, the pupils should write the decimal point in the quotient when they reach it in the dividend, placing it under the latter. In long division, the decimal point in the quotient is placed over the point in the dividend. While the scholars have been warned in their early work in short division against writing 02 as the quotient of 8)16, they will see the need of the prefixed cipher in the answer, .02, to 8).16. From the inspection of a few examples of this kind, they will understand that each figure of the dividend after the decimal point requires a quotient figure (or cipher).

In the long-division examples worked out in Art. 563 of the Arithmetic, the partial products are omitted, to show how some European countries shorten work of this kind. The horizontal lines given in the text-book are not used.

```
            1.875
         64)120.
            560
            480
            320
              0
```

After the pupil writes the quotient figure 1, he subtracts by the "building-up" method. The second remainder, 48, is obtained by saying 8 fours are 32, and 8 (writing it) are 40; 8 sixes are 48, and 4 (carried) are 52, and 4 (writing it) are 56. See Arithmetic, Art. 385.

564. Teachers should not forget that systematic instruction in decimal fractions belongs to the sixth school year. They should be content if their pupils learn to place correctly the decimal point in the quotient.

Nos. 21–30 should be considered rather as examples in division, than as examples in the reduction of common fractions to decimal ones. By $\frac{1}{4}$ is meant $1 \div 4$; and to solve it, the pupil may write 4)1.00, as in No. 11. He should learn by degrees, however, that it is not necessary to write all the ciphers in the dividend in order to obtain the result. The answer to No. 22 can be derived by a bright scholar from 8)1.0 just as well as from 8)1.000. He may desire the first cipher as a starting-point, but he finds the others superfluous.

565. These problems contain a few simple applications of the decimal work learned thus far. In No. 1, the value of the franc may be given in the fractional form also, $19\frac{3}{10}$¢, and the problem worked fractionally and decimally.

Approximate sight results might be asked before written work is begun. Taking the franc as about 20¢, or $\$\frac{1}{5}$, the pupil should say that the answer to No. 1 is less than $250. Assuming 40 inches, or 1$\frac{1}{9}$ yd. as the length of the meter, 1800 meters would be equal to 2000 yards. No. 33 can be solved without using the pencil. In No. 38, the first result may be in the form of a common fraction, $\frac{1}{8}$ peck, to be changed, as in No. 22, to .125 peck.

569. The remarks made in Art. 564 as to the formal teaching of decimal fractions is equally applicable to the subject of measurements. At one time, all instruction in mensuration was deferred until the last year of the common-school course. At present, this subject is generally taken up in connection with the systematic work in denominate numbers; but there is no good reason why pupils compelled to leave school by the end of the fifth year of the course should not receive so much practice in finding the areas of rectangles as they have time for and can readily understand.

In most city schools, the children of this grade know from their lessons in form and drawing what is meant by a square and a rectangle. If pupils are not familiar with these terms, they should be explained.

That scholars may obtain a good idea of a square inch, they should be required to cut out a number of square pieces of paper, each side measuring an inch. These squares should then be used in determining the number of square inches in the two rectangles next drawn; 2 inches by 1 inch, and 3 inches by 2 inches. Children that determine the areas of these rectangles by covering them with their paper squares will have a better knowledge of 2 square inches and 6 square inches than if they merely divide the rectangles by lines as suggested in the textbook.

The larger rectangles, 6×3 and 4×4, may be cut up into inch squares by drawing lines, and the rule for obtaining the area of each deduced from an examination of the figures. In the first, the pupil will see that he has 3 rows of squares, 6 to a row (or 6 rows, 3 to a row), making 6×3, or 18, squares. The rule should take in his mind some such form as this: the number of square inches in a rectangle is equal to the product of the number of inches in one dimension multiplied by the number of inches in the other; but he should not be required to give it expression. The teacher should take care that he does not think that "inches by inches give square inches."

570. As has already been said, the chief use, in arithmetical instruction, of objects, diagrams, etc., is to enable pupils to work without them. After the scholars understand how to obtain the area of a rectangle, they should cease to draw the figure and to subdivide it into squares.

It will be noted that the answers to the first 20 examples are to be given in square inches. In Nos. 11–20 each dimension should be reduced to inches before the multiplication is performed.

X

NOTES ON CHAPTER SEVEN

At this point regular fraction work should begin. From time to time, as occasion offers, the meanings of the technical terms should be elicited from the pupils; but the teacher should neither accept a memorized definition that is not thoroughly understood, nor should she require absolute correctness in the phraseology of a definition made by a scholar.

577. While the denominators of the fractions should generally be small; and while common denominators should, as a rule, be determinable by inspection, it is necessary, nevertheless, that the children be taught how to handle such other fractions as they may occasionally meet. It is not necessary that they should grasp the exact meaning of $\frac{287}{384}$ in the answer to No. 4, although proper teaching may enable them to see later on that this fraction approximates $\frac{288}{384}$, $\frac{3}{4}$.

In these earlier examples, the inspection method of determining the common denominator is continued.

580. Besides being necessary as a preliminary to subsequent work in fractions, expertness in determining the factors of a number is useful in enabling pupils to shorten their work by cancellation. The teacher should use these and similar exercises again and again, for a few minutes at a time, until her scholars can give the answers with great rapidity.

581. The pupils will need to learn the difference between the three factors of 12, and three divisors of 12. The factors will be 2, 2, and 3, because their product, $2 \times 2 \times 3$ equals 12. The divisors of 12 are 2, 3, 4, and 6.

These exercises, as well as those in Arts. 582–585, are not so valuable as to demand the reviews suggested for those in Art. 580.

In finding three (or more) factors of a number, the scholar should commence with the smallest. The first of the three factors of 8 is 2; dividing 8 by this factor, 4 is obtained, of which 2 and 2 are the factors.

The three factors of 18 are $2 \times 3 \times 3$; of 20, are $2 \times 2 \times 5$; of 27, $3 \times 3 \times 3$; etc.; etc.

582. It is customary to define a prime number as one that has no factor except itself and unity. The omission of the last four words will not mislead any person, as there could be no prime numbers if 1 were considered a factor. When the factors of a number, say 20, are asked for, no one gives $1 \times 1 \times 1 \times 1 \times 1 \times 2 \times 2 \times 5$ as the answer, or says that 20 has eight (or more) factors.

588. In reducing these fractions to lowest terms, it is not necessary that the pupils should use the greatest common divisor. See Arithmetic, Art. 592. On the other hand, they should not waste time in dividing each term by 5, if 25 is a common divisor.

589. Pupils should not be permitted to forget these tests of the divisibility of numbers. To those given in the text-book, there may be added that when a number divisible by 3 is even, it is also divisible by 6.

While a teacher should know that 1001, with, of course, its multiples, — 2002, 6006, 15015, etc., — is divisible by 7, 11, and 13, she should not burden her scholars with the information; nor should she dwell upon the test of divisibility by 8.

591. Beginners should be taught only one method of finding the greatest common divisor, and the one here given is applicable to all kinds of numbers. Teachers should not bewilder young pupils by endeavoring to make them understand the principles upon which this method is based.

595. Many teachers prefer to permit their pupils to write down all of the denominators, and then to strike out any one that is repeated or that is the factor of any other. They think the pupil is less likely to make a mistake by following this plan. In no case should scholars be permitted to begin work before rejecting or striking out the unnecessary numbers.

$$2\ \overline{\smash{\big|}\ \cancel{3}\text{-}9\text{-}\cancel{7}\text{-}\cancel{14}\text{-}\cancel{6}\text{-}14\text{-}\cancel{2}\text{-}12}$$
$$976$$
etc.

605–606. Pupils that have had regular drills in the combinations given in the previous chapters will be able to take the extra step required by these examples. See Arts. 286–290 and 350–352.

607. A scholar that can find mentally the cost of 47 articles at 25¢ each should be able to give the product of 47×25 or 25×36 without using the pencil, and the teacher should give him a chance to determine for himself the method of doing it.

609. Such questions as $18\frac{2}{3} \div 2\frac{2}{3}$ can be worked by the method given in the last chapter; viz., 56 thirds ÷ 8 thirds = 56 ÷ 8 = 7. Those contained in the 4th column should not be used until the pupils have had formal instruction in division of fractions. When they are taken up, the method followed should be that given above, the fractions being reduced to a common denominator, etc. $\frac{1}{2} + \frac{2}{3} = 3$ sixths + 4 sixths $= 3 + 4 = \frac{7}{6}$; $\frac{2}{3} + \frac{3}{4} = 8$ twelfths + 9 twelfths $= \frac{17}{12}$; etc. See Arithmetic, Art. 639, note.

610. No. 9: For $1.25 I can buy 5 times as many pounds as for 25¢, or 15 pounds. No. 16: For 18¢ there can be bought $\frac{18}{24}$ lb., or $\frac{3}{4}$ lb., or 12 oz.

613. In giving answers to these exercises, pupils should be permitted to write the fraction first and then the whole number.

The object of these exercises is to accustom the scholars to dispense with writing unnecessary reductions in adding and subtracting simple fractions.

614. Another method of finding the difference between $11\frac{3}{8}$ and $6\frac{3}{4}$ is to take $6\frac{3}{4}$ from 7, obtaining $\frac{1}{4}$, and to add to this the difference between 7 and $11\frac{3}{8}$, or $4\frac{3}{8}$.

In some classes of sight exercises, those given in Arts. 587, 594, and 605–609, for instance, the pupils should not take pens to write the result until told to do so by the teacher, after sufficient time has been given to obtain the answer. In the exercises of Arts. 613 and 614, the pupils should be permitted to take their pens at once, and to write each part of the result as soon as it has been obtained. Arts. 584, 650–654, 699, etc., also contain exercises of this kind.

616. See Art. 563. The first quotient figure, 4, is written. The pupil then says 4 sixes are 24 and **4** (writing it) are 28; 4 threes are 12, and 2 (to carry) are 14, and **3** (writing it) are 17. This gives the first remainder, 34. The next figure, 3, is then brought down, and 9 is written in the quotient.

$$495\tfrac{17}{36}$$
$$36)17837$$
$$343$$
$$197$$
$$17$$

The product of 9 times 36 is subtracted from 343 as given above, to obtain the next remainder, 19. The pupil says 9 sixes are 54 and **9** (writing it) are 63; 9 threes are 27 and 6 (to carry) are 33, and **1** (writing it) are 34.

618. While pupils should be encouraged to shorten their work by cancellation, the slower children should not be censured when they overlook some cases in which it is possible to employ this expedient. In these examples, however, all the scholars should be required to indicate the operations, and then to cancel.

619. It is not supposed that pupils should write answers to these questions, as is generally done in the case of the oral problems. These exercises are intended to lead up to the rule for multiplication of fractions. Diagrams should be drawn on the blackboard by the pupils, to illustrate the answers, but the teacher should refrain as much as possible from "explaining."

The board work should be done chiefly by the more backward members of the class rather than by the brighter ones. In illustrating fractions by diagrams, the unit employed should generally be a circle, the part dealt with being distinguished by shading.

The zealous teacher should not become discouraged at the inability of some members of the class to thoroughly grasp the mathematical principles involved in this and other operations. Even the ability to handle fractions mechanically will be of great use in after life, and all the pupils can be taught at least this much.

624. No. 31 reduces to $9\frac{3}{4} \div 3$, which can readily be worked by the pupils without assistance. No. 34, $40\frac{1}{2} \div 8$, may be difficult for some; but the teacher should not offer help too soon. The second term of No. 36 is easily obtained; and No. 40 will give no trouble.

625. These exercises are to be used in the same way as those in Art. 619.

Many children find it difficult to understand that 4 fourths ÷ 3 fourths $= 1\frac{1}{3}$. They think that the answer should be $1\frac{1}{4}$, reasoning it out in some such way as this: 3 fourths into 4 fourths goes 1 time and 1 fourth over. They fail to recollect that the remainder in division is written over the divisor, which would give them $1\frac{1 \text{ fourth}}{3 \text{ fourths}}$, or $1\frac{1}{3}$. If they have been well taught previously, they may remember that 4 fourths ÷ 3 fourths $= 4 \div 3 = 1\frac{1}{3}$. Even a fairly bright pupil, when asked how much tea at $\$\frac{3}{4}$ per lb. can be bought for $\$1$, will sometimes reply, "A pound and a quarter." When he is told that his answer is correct if he means by it a pound of tea and a silver quarter, he sees the mistake and changes the result to "a pound and a third."

626. If $1 \div \frac{3}{4}$ is made concrete, a pupil can more easily show by a diagram that the result is $1\frac{1}{3}$. A problem of this kind may

NOTES ON CHAPTER SEVEN

be given: If it requires $\frac{2}{3}$ yd. of material for an apron, how many aprons can be made from 1 yard? A rectangle is drawn to represent the yard of material, and it is divided into thirds. Underneath, a rectangle two-thirds as long is drawn to represent the quantity required for an apron. When the pupil compares the two rectangles, he sees that the portion remaining after one apron is made will supply sufficient material for one-half of another.

627. While it is generally better in oral work to divide one fraction by another by reducing both to a common denominator, it will be found simpler in written work to have pupils invert the divisor.

631. "Invert the divisor, and proceed as in multiplication," is the rule generally followed.

634. No. 33 can be shortened by writing it as follows, before beginning work: $\frac{20}{1} \times \frac{3}{4} \times \frac{4}{5}$, the divisor being inverted. No. 34 should be treated in the same way: $(20 \div \frac{7}{8}) \times \frac{3}{4} = \frac{20}{1} \times \frac{8}{7} \times \frac{3}{4}$. The divisor of No. 35 consists of two fractions, both of which should be inverted: $20 \div (\frac{7}{8} \times \frac{3}{4}) = \frac{20}{1} \times \frac{8}{7} \times \frac{4}{3}$. This method should be followed with Nos. 36, 41, and 42. The first of these becomes $\frac{20}{1} \times \frac{8}{7} \times \frac{4}{3}$; the next, $\frac{17}{8} \times \frac{3}{4} \times \frac{28}{5} \times \frac{5}{28} \times \frac{3}{4}$; and the next,

$$\frac{11}{2} \times \frac{81}{4} \times \frac{19}{6} \times \frac{25}{4} \times \frac{4}{11} \times \frac{5}{21} \times \frac{1}{81}.$$

635. Each teacher must determine for herself what method of analysis should be encouraged in such questions as Nos. 4, 8, 13, and 14. While set forms should be avoided, children need direction in the solution of problems of this kind.

In solving No. 4, for instance, the greater number of teachers prefer to have pupils first find the cost of $\frac{1}{2}$ yd. When this method is followed, care must be taken that all the pupils understand why $\frac{1}{2}$ yd. costs one-half of 20 cents. This may be made clearer to some by writing the fractions in this way: If 2 thirds (or parts) cost 20 cents, what will 1 third (or part) cost? A diagram similar to that given in Arithmetic, Art. 636, may help others to understand the method.

Other successful teachers think the written work is benefited by treating these examples as problems in division. They lead their children to determine in each case what operation is involved, by requiring them to consider what they would do if the fraction were a whole number. In No. 1, for example, the cost of 16 balls at \$3 each would be \$3 × 16. In No. 2, the pupil would say, "If I paid \$12 for base-balls at \$3 each, the number of balls would equal 12 ÷ 3. I must, therefore, divide." He mentally inverts the divisor, $\frac{3}{4}$, then cancels, etc.

636. The scholars should be allowed sufficient time to work these out in their own way.

639. No. 4: $24\frac{1}{2} \div 3\frac{1}{2} = \frac{49}{2} \div \frac{7}{2} = 49 \div 7$. Some pupils will see that time is lost in No. 6 by finding the cost of a pound. No. 7 is an example in division: $1\frac{2}{3} \div 2\frac{1}{2} = \frac{5}{3} \div \frac{5}{2} = \frac{5}{3} \times \frac{2}{5}$; or $\frac{10}{6} \div \frac{15}{6} = 10 \div 15$, etc. In No. 16, 36 hats will cost 3 times \$7.

642. See Art. 546.

649. In multiplying by 25, the pupil is generally told to annex two ciphers and to divide by 4. In mental work especially, the annexation of the ciphers confuses some scholars by giving them a larger dividend than is really required. The product of 25 times 19 may be obtained more easily by taking one-fourth of 19, or $4\frac{3}{4}$, and changing this quotient to 475, than by finding one-fourth of 1900. In No. 9, the pupil should see that at \$100 per bbl., the pork would cost \$5600, and that at \$12.50 per bbl. ($\frac{1}{8}$ of \$100), it would cost $\frac{1}{8}$ of \$5600.

650. In No. 1, divide 837 by 4, and for the 1 remainder affix 25 to the quotient. In No. 4, annex two ciphers to the quotient of 508 by 4. In No. 9, affix 250 to the quotient of 837 ÷ 4.

In multiplying 6281 by $12\frac{1}{2}$, No. 18, divide 6281 by 8, obtaining 785, and annex $12\frac{1}{2}$ for the 1 remainder, making the result $78,512\frac{1}{2}$.

654. When the divisor is a whole number, time should not be wasted in changing a mixed number dividend to an improper fraction. Nos. 64, 65, and 66 resemble those already worked. In No. 69, after obtaining the quotient 14, there will be a remainder $2\frac{1}{2}$, which is changed to $\frac{5}{2}$ and divided by 5, giving $\frac{1}{2}$ as the result. In No. 69, the remainder, $5\frac{1}{4}$, is changed to $\frac{21}{4}$, which gives $\frac{21}{32}$ when it is divided by 8.

656. Some mistakes would be avoided if pupils would learn to ask themselves if the answer they have obtained is a reasonable one. Permit the scholars to work out all these examples without giving them a rule for " pointing off."

669. See Art. 521.

- 3. 6×6.
- 4. $300 \div 12$.
- 5. 86×1.
- 6. $36 \div 4$.
- 7. $800 \div 100$.
- 8. 8×8.
- 9. 7×11.
- 10. $64 \times \frac{3}{8}$.

670.

- 3. 25×12.
- 4. $36 \div 6$.
- 5. $86 \div 1$.
- 6. 32×5.
- 7. $800 \times .1$.
- 8. 8×8.
- 9. 7×12.
- 10. $64 \div \frac{1}{4}$.

671.

- 2. $\$2\frac{1}{2} \times 200$.
- 3. $25¢ \times 4$.
- 4. $\$12 \times 400$.
- 5. $\$2 \times 8$.
- 6. $25¢ \times 800$.
- 7. $\$2\frac{1}{2} \times 20$.
- 8. $60¢ \times 1000$.
- 9. $\$5000 \times 7$.
- 10. $\$1 \times 6$.

677. Teachers should carefully avoid giving unnecessary "rules." There is no good reason why an average pupil should not be able to determine for himself how to ascertain what part of $15 a man has spent when he has spent $5. While the introduction of fractions into such an example makes it more difficult for the scholar to give the answer off-hand, his instruction up to this time should have taught him that the same process is to be employed. A pupil should be required to depend upon himself to at least a reasonable extent.

678. As a preliminary to the work in denominate numbers in the next three pages, the teacher should place on the board a few such examples as the following, to which the scholars should give answers at sight:

$$1\tfrac{1}{2} \text{ qt.} \qquad 1 \text{ qt. } 1 \text{ pt.} \qquad 3 \text{ qt.} \qquad 1 \text{ qt. } 1 \text{ pt.}$$
$$+1\tfrac{1}{2} \text{ qt.} \qquad +1 \text{ qt. } 1 \text{ pt.} \qquad -1 \text{ qt. } 1 \text{ pt.} \qquad \times 2$$

$$2\overline{)3 \text{ qt.}} \qquad \qquad 1 \text{ qt. } 1 \text{ pt.}\overline{)3 \text{ qt.}}$$

Nearly every member of the class will be able to obtain the results in a moment, without any suggestions from the teacher. If the examples are left on the board, the pupils can refer to them for aid in working some of those found in the text-book.

The teacher that wishes to develop power in her scholars should be careful not to give a particle more assistance than is necessary. She should permit the children to deduce from the above examples the rules necessary to solve the others, being patient if the pupils are somewhat slow in doing this work. When, however, a circuitous method has been employed, she should lead the class to see how the work can be improved by the use of a shorter way.

680. It may be necessary to take up again, for purposes of review, the preliminary exercises of the previous chapter. See Art. 569, pp. 56 and 57.

NOTES ON CHAPTER SEVEN

681. As the table of square measure is not introduced until the next chapter, it will be necessary to reduce to yards the dimensions that are given in feet or inches.

2. 18 yd. by 21 yd.
3. 2 yd. by 3 yd.
7. 9 yd. by 32 yd.
8. 18 yd. by 2 yd.
9. 16 yd. by 15 yd.
10. 1½ yd. by 24 yd.

682. No. 14. 14 yd. by ⅔ yd.

No. 15. 8 pieces, each 36 yd. long and $\frac{27}{44}$ yd., or ⅝ yd. wide.

No. 18. See Arithmetic, Art. 818, problem 20. A modification of this diagram, showing four squares instead of four rectangles will be the drawing required, except that the squares above and below need not necessarily occupy the positions there indicated.

XI

NOTES ON CHAPTER EIGHT

With this chapter begins the regular work in decimal fractions, and the pupils should now be taught the principles underlying the various operations.

685. While pupils may know that $\frac{23}{8}$ means that 23 is to be divided by 8, it may be well to lead them again to see that $\frac{3}{4}$ is the same as $3 \div 4$, or $\frac{1}{4}$ of 3. After they understand that every common fraction may be considered an "indicated division," they will understand that the decimal fraction obtained by performing this operation is the equivalent of the common fraction whose denominator is used as a divisor and whose numerator is used as a dividend. See Arts. 563 and 564.

686. As previous work in decimals has been confined chiefly to three places, some review and extension of the notation and numeration exercises of Arts. 547–551 may be necessary.

687. After writing each of these decimals in the form of a common fraction, a scholar should be able to determine at a glance whether or not it can be reduced to lower terms. This reduction is possible when the decimal is an even number or terminates in a 5.

While it is inadvisable to waste time in calculating the greatest common divisor, pupils should be encouraged to use large divisors; 4 rather than 2, when possible, and 25 rather than 5.

688. The common fractions contained in these exercises are such as do not require much calculating to change them to decimals. The scholars should be able to write the numbers in vertical columns directly from the text-book, making the necessary reductions mentally.

In reducing $\frac{1}{16}$ to a decimal, it may be easier for some to consider it $\frac{1}{4}$ of $\frac{1}{4}$, or $\frac{1}{4}$ of .25. The reduction of $\frac{23}{50}$ is simplified by multiplying each term by 2, making it $\frac{46}{100}$, or .46, instead of dividing 23 by 50, etc., etc.

690. Nos. 62, 64, 66, and 68 may be worked by using the common fraction given, and also by reducing this to a decimal before performing the multiplication.

691. See Art. 563, p. 55, and Art. 616.

692. The teacher should not permit the employment of long division in these examples. In No. 92, the children can see that changing the dividend to .18756 divides it by 100, and that .18756 ÷ 3 is the same as 18.756 by 300. See Arithmetic, Art. 668.

694. Ciphers at the right of a decimal should be rejected, excepting, perhaps, the final 0 in cents. See Nos. 3, 4, and 10.

2. $ .95 × 7.6.
3. $ 2.80 × 48.6.
4. $ 21.30 × 39.25.
5. $ .68 × 18.75.
6. $ 22 × 108.745.
7. $ .75 × 148.6.
8. $ .13¼ × (2376 ÷ 12).
9. $ 35 × 4.5.
10. $ 13.50 × [(28 × 12) ÷ 144].

While it is inadvisable to confuse children by too many short methods in the earlier stages, they should be encouraged in examples like the foregoing to use as a multiplier the number that will make the work easier, and to employ a common fraction instead of a decimal whenever the use of the former would lighten their labor. In No. 1, for instance, the result is obtained with fewer figures by multiplying 24.4 by 6¼, instead of 6.25 × 24.4.

695. The operation should first be indicated.

11. $\dfrac{\$.90 \times 38648}{60}$.

12. $\dfrac{\$5 \times 18964}{2000}$.

13. $\dfrac{\$.36 \times 48576}{32}$.

14. $\dfrac{\$1.83 \times 69104}{2 \times 56}$.

etc., etc.

703–704. See notes on previous special drills, Arts. 286 and 350.

705. See Arts. 528 and 649. In multiplying 46 by $33\frac{1}{3}$, divide 46 by 3, which gives $15\frac{1}{3}$, and substitute $33\frac{1}{3}$ for the fraction in the quotient, thus obtaining the result, $1533\frac{1}{3}$.

706. $975 \div 25 = 9\frac{3}{4} \div \frac{1}{4} = 9\frac{3}{4} \times 4.$ $433\frac{1}{3} \div 33\frac{1}{3} = 4\frac{1}{3} \div \frac{1}{3} = 4\frac{1}{3} \times 3.$

708. Use first as "sight" problems, if the pupils find the numbers too large to be carried in the mind. By degrees, however, they should acquire the power to solve problems of this kind without seeing the figures, especially when the operations are not numerous or involved.

709. In such examples as Nos. 1, 9, 10, 11, and the like, many children fail to comprehend the form of analysis generally given. While they get some facility in applying the method, they do not understand the underlying principle. In finding a number, $\frac{5}{6}$ of which is 180, they learn to divide by 5 and to multiply the quotient by 6, and to repeat the customary formula, without knowing the reasons for the different operations. There are only four fundamental processes in arithmetic, and children should be taught to determine for themselves which to use in a given example that is within their experience, rather than to depend upon a rule which they do not fully understand, and which they are likely to forget or to misapply. See Art. 635. A few diagrams are here introduced, to be used by the teacher that does

NOTES ON CHAPTER EIGHT 71

not wish her pupils to obtain in No. 1, for instance, the length of the room by dividing 15 by ⅚. The five spaces in the width are each 3 ft., which will make the length 18 ft. When a scholar understands this from the diagram, he can understand that when ⅚ of a number is 15, ⅙ is 15 ÷ 5, or 3.

While, for purposes of drill, many "abstract" examples of the same kind are brought together in one place, care has been taken in the problems to avoid having two consecutive ones alike in character. Problem work to be of value should not be permitted to become mechanical. Pupils should need to study each problem to determine the method of solving it.

710. See Arithmetic, Art. 384.

716. By placing the multiplier at the right of the multiplicand, the pupil can use the latter as the first partial product, instead of writing it again, as he would be compelled to do if the multiplier were placed in its usual position.

717. Some teachers might prefer to place the product by 8 above the multiplicand, as being the form to which the scholars are more accustomed; but in such an example as No. 26, the latter part of the work can be shortened only by placing the product by the units' figure under the other.

$$576 \times 14 \times 21$$
$$2304$$
$$8064$$
$$16128$$
$$\overline{169344} \; Ans.$$

$$19744$$
$$2468 \times 18$$
$$\overline{44424} \; Ans.$$

No. 28 should be worked as is here shown. Annexing two ciphers to the multiplicand and multiplying by ⅔ gives the product by 66⅔.

$$48600$$
$$\underline{2}$$
$$3)\overline{97200}$$
$$32400$$
$$226800$$
$$\overline{2300400}$$

719–720. See Art. 521.
 1. 24 lb. @ $¼.   3. 64 yd. @ $⅞.
 2. 24 horses @ $125.   4. 485 bu. @ $1.

5. 96 lb. @ $\$\frac{1}{4}$.
6. 840 yd. @ $\$\frac{1}{3}$.
7. 360 yd. @ $\$\frac{2}{3}$.
8. 48 cwt. @ $\$\frac{5}{8}$.
9. 92 hats @ $\$1\frac{1}{2}$.
10. 128 lb. @ $\$\frac{1}{8}$.
11. $27 \div \frac{1}{4}$.
12. $300 \div 1\frac{1}{2}$.
13. $24 \div \frac{3}{8}$.
14. $15 \div \frac{1}{8}$.
15. $60 \div 2\frac{1}{2}$.
16. $32 \div \frac{1}{3}$.
17. $70 \div \frac{7}{8}$.
18. $60 \div \frac{5}{8}$.
19. $64 \div \frac{2}{3}$.
20. $28 \div 1\frac{3}{4}$.
21. 17×4.
22. $256 \times \frac{1}{4}$.
23. 25×16.
24. 6×6.
25. 86×1.
26. 33×5.
27. $800 \times \frac{1}{8}$.
28. 8×8.
29. 7×11.
30. $64 \times \frac{3}{8}$.

721. The decimals should be reduced to common fractions whenever the work is rendered easier by the change.

1. $360 \times \frac{1}{4}$.	7. $72 \times \frac{1}{4}$.	13. $84 \times \frac{1}{4}$.
2. $560 \times \frac{1}{8}$.	8. $84 \times \frac{3}{4}$.	14. 15×6.
3. $240 \times \frac{3}{8}$.	9. $96 \times \frac{1}{20}$.	15. 4×4.
etc.	etc.	etc.

722. The pupil should employ such method as is best adapted to the particular example:

1. $240 \div \frac{1}{2}$.	9. $48 \div \frac{1}{200}$.	17. $65 \div \frac{1}{8}$.
2. $360 \div \frac{3}{4}$.	10. $72 \div \frac{1}{40}$.	18. $840 \div 8$.
3. $45 \div \frac{1}{8}$.	11. $92000 \div 2$.	19. $11 \div \frac{1}{16}$.
etc.	etc.	etc.

723. Nos. 1 to 8 are intended to furnish practice in sight cancellation. In Nos. 13 to 16, the reduction of the multiplier to an improper fraction will simplify the work for some pupils.

726. Whenever possible, the least common denominator should be determined by inspection.

735. Do not give "rules." See Art. 678.

736-737. These exercises are introduced to accustom the pupils to add and to subtract simple mixed numbers without rewriting the fractions reduced to a common denominator.

740. In these and other similar examples, the teacher should not anticipate the work of the higher grades by systematic instruction in advanced topics. All that should be done with respect to these problems is to show the pupils that, when a solution involves multiplication and division, time may frequently be saved by means of cancellation. The pupils should be permitted to work out No. 1 at length, if they wish; after which they should be required to indicate the work by signs, and then to cancel. Division should, of course, be indicated by writing the divisor as a denominator.

Some excellent teachers require their scholars before beginning work on a problem to indicate by signs all the operations necessary to its solution, thereby compelling them to study the conditions thoroughly at the outset. Too many pupils commence to add, subtract, etc., without fully realizing what is required in a given example.

742-744. See Art. 678.

745. The pupils should write the dimensions on each diagram, changing them, when necessary, to the denomination required in the answer.

746. The formal study of percentage belongs to the next year of the course, and teachers should not dwell too much on this topic. After the pupils understand the meaning of the term *per cent*, they should be able to work the examples given. Other technical terms, definitions, etc., should be omitted for the present.

753. The pupils will readily see that the words " Bought of," used in Arts. 546 and 642, are inappropriate in bills for work done. No. 5 may be made out in the form here shown or similar to the one given in Art. 546. See Art. 642 for a bill for goods bought at different times; or use the heading given in this article.

754. What has been said about percentage in Art. 746, is applicable to this topic. Such children as hear their parents talk of savings-banks, etc., know sufficient about interest for the purposes of this chapter. No rules should be given.

756. The pupils should deduce their own rule for calculating the area of a right-angled triangle.

758. In Art. 653 the pupils have been taught to multiply $18\frac{2}{3}$ by 6 in one line; in Art. 654, they have learned how to divide $18\frac{2}{3}$ by 2, which is the same as finding $\frac{1}{2}$ of $18\frac{2}{3}$, so that nothing new is here presented.

763–764. Although these examples are not strictly practical, they are useful in giving the pupils the facility necessary to perform readily operations involving fractions or decimals. While it is not necessary to work them all, the scholars should by this time have acquired such expertness in the fundamental operations as to be able to obtain the results in a very short time.

765. See Arithmetic, Art. 591.

XII

NOTES ON CHAPTER NINE

The technical terms used in denominate number work should now be regularly employed by teacher and pupil, but set definitions should not be memorized. The scholars should be required to arrange their work properly, and to perform the various operations with as few figures as are consistent with accuracy.

767. In reducing 16 gal. 1 qt. to quarts, the pupil should write 65 qt. at once. He multiplies by 4, saying 4 sixes are 24, and 1 are 25 — writing the 5, etc. In reducing $31\frac{1}{2}$ gal. to quarts, the work should occupy but a single line. See Arithmetic, Art. 653.

770. No special rule should be given in Nos. 33, 34, and 35 for the reduction of a fractional or a decimal denominate unit.

773. A pupil should be permitted to work such examples as No. 2 in his own way. They do not occur frequently enough in practice to make it advisable to give them special treatment; but the teacher should suggest, as in other exercises, the advisability of shortening the work by indicating operations and cancelling. Thus,

$$12 \text{ min. } 30 \text{ sec.} = 12\frac{1}{2} \text{ min.} = \frac{12\frac{1}{2}}{60} \text{ hr.} = \frac{12\frac{1}{2}}{60 \times 24} \text{ da., etc.}$$

5. $\dfrac{9}{60 \times 24}$ da.

6. 750 lb. $= \frac{750}{2000}$ T. $= \frac{3}{8}$ T.; $\$5 \times 5\frac{3}{8} = \$26.87\frac{1}{2}$, or $\$26.88$.

Ans.

7. No. of tons = \$18.76 ÷ \$5 = 3.752; .752 T. = (.752 × 2000) lb. = 1504 lb. *Ans.* 3 T. 1504 lb.

8. 7 T. 296 lb. = 14296 lb.; (\$35.74 ÷ 14296) × 18748 = *Ans.*

9. 9 T. 1568 lb. = 19568 lb.; \$48.92 ÷ 19568 = cost per lb. \$73.11 ÷ (\$48.92 ÷ 19568), or (\$73.11 × 19568) ÷ \$48.92 = number of pounds. Reduce to tons, etc.

774. By this time, the pupils should know how to add compound numbers, so that the chief duty of the teacher should be to see that the operation is not spun out too much. A scholar of this grade should not find the total number of ounces in **1** by adding each column separately; he should say 27, 36, 39 oz., or 2 lb. 7 oz., without writing anything but the 7 oz., which is put in its proper column and 2 lb. carried.

In **4**, the addition of the units' column of minutes gives a sum of 15. Since minutes are changed to hours by dividing by 60, which ends in a cipher, the units' figure of the remainder will be 5, so that this figure may be written in the total. Carrying one, the sum of the tens' column is 11, which contains 6 once with a remainder of 5. This is written in its place, making 55 minutes, and 1 hour is carried. The two columns of hours are added in one operation—21, 38, 43, or 1 day 19 hours. **6** should be treated in the same way, no side work being permitted.

In **7**, the pounds are reduced to tons by dividing by 2000, so that the sum of the units', tens', and hundreds' columns of pounds may be written in the total, the sum of the thousands' column being divided by 2 to reduce to tons.

775. Nothing should be written but the results. In **27**, the addition of 1 ton to 1552 lb. will change only one figure of the latter, and this change can be carried in the head. In **29**, 320 rods should be added to 15 rods mentally and 24 rods deducted from the sum, only the answer being written.

779. In dividing 5 bu. by 4, **79**, the answer is not to be given as $1\frac{1}{4}$ bu.; the division should be continued through pecks. The result in **88** should contain weeks, days, hours, and minutes.

784. While these drills seem somewhat difficult for mental work, they should not be too severe for children that have been studying arithmetic for over five years, especially if the previous drills have been faithfully attended to. The ability of many children to handle numbers seems to decrease after the fourth school year, the greater portion of the subsequent instruction being given to new topics to the neglect of continued practice in the fundamental processes. The conscientious teacher should remember that the bulk of the mathematical work of most of her scholars after they leave school will not extend much beyond what has been learned in the first four years.

The ability to handle at sight or mentally such numbers as are here given, will be of use to the scholars in various ways. The average pupil attends to only one figure at a time; and he is frequently unable, after a simple addition or multiplication, to see that his answer is very far astray. Practice with such drills as these, and in the sight approximations, will enable him to test his work in such a way as to detect any very serious error.

Scholars find it easier to add or subtract such numbers as 163, 8610, etc., when they are read "one, sixty-three;" "eighty-six, ten;" etc. Following the order in which the figures are read seems the most natural way in mental work. When a pupil is asked to find the sum of 163 and 137, he is less likely to make mistakes if he proceeds in this way: 263, 293, 300; adding to the first number — 163 — 100, 30, and 7 in the order in which the figures are repeated to him.

786. In multiplying 21 by 15, 41 by 14, etc., the scholar generally finds it easier to commence with the tens: 15 twenties are 300, 15 ones are 15 — 315; 14 forties are 560, and 14 are 574.

$48 \times 16\frac{2}{3}$ becomes $\frac{1}{6}$ of 48 hundred; $32 \times 37\frac{1}{2} = \frac{3}{8}$ of 32 hundred, etc.

787. These exercises present rather more difficulty, and are probably not so useful, on the whole, as the others. For this reason, they should be employed as sight work chiefly.

788. In 13¾×5, multiply 13 first by 5, and then ¾, obtaining 65+3¾, or 68¾. In dividing 24 by 2⅔, reduce both to thirds — 72 thirds ÷ 8 thirds = 72 ÷ 8 = 9.

790. The teacher should not neglect such addition exercises as are scattered throughout the book.

791. It happens occasionally in multiplying by a mixed number, that the units' figure of the integer and the numerator of the fraction are the same. In such a case, a few figures will be saved by following the method given in the text-book, instead of writing again the product by 3 as shown above.

```
     4846
    × 3⅗
   ───────
   5)14538
     2907⅗
    14538
   ───────
      etc.
```

792. The product by 100 may be placed above the number, if desired. In multiplying by 1000, the multiplicand is subtracted from 1000 times itself. To find the product of 9832 by 990, multiply by 99, and annex a cipher to the result. Taking one-fourth of 268400 gives the answer to **21**.

```
2761 × 999
2761000
2758239  Ans.
```

800. The pupils should find for themselves in **5** the number of square inches in a square foot, etc. A drawing is asked in the first part of **14**, so that children will see that the dimensions are not 4 × 6. The short method of finding the area of the fence in **15**, by multiplying 900 by 10, should not be given yet: the scholars should be permitted, for the present, to calculate the area of one part at a time. In **16**, it is suggested that the area of the walk be ascertained by subtracting from the whole area (250 × 200) sq. ft., the area of the part left for the garden (230 × 180) sq. ft.; but the scholars should be encouraged to calculate the surface of the walk in another way, such as by taking the two ends as measuring each 250 ft. by 10 ft., and the sides as 180 ft. each by 10 ft. The number of square feet in the sidewalk of **17** will be (270 × 220) − (250 × 200); or (270 × 10) + (270 × 10) + (200

NOTES ON CHAPTER NINE 79

× 10) + (200 × 10). For **20**, a modification of the diagram in Problem 20, Art. 818, is desired.

801. To show pupils what is required in **21**, a pasteboard box, without a cover, may be opened out as is represented in Problem 2, Art. 871, the upper rectangle (the bottom of the box) representing the ceiling.

802. 3. The sixth dose will be taken at 7 o'clock, the second at 3 o'clock, the fourth at 5 o'clock. **4.** He works 6 days. **6.** A fence 6 ft. long will require 2 posts; a 12-foot fence will require 3 posts; a fence 120 ft. long will require 21 posts.

803. In finding the time between two dates, the first date is excluded except when the contrary is expressly specified.

804. 11. 30 days + 19 days. **12.** 0 days in October + 30 days in November + 30 days in December.

805. 1. In February, there are (29 − 6) days, or 23 days. **3** and **4.** Leap year. **11.** Jan. 8, 15, 22, 29; Feb. 5, 12, 19, 26 are Sundays. The man works 30 days in January and 28 in February, less 8 Sundays and 1 holiday.

807–809. See Arts. 746 and 754.

808. 3–13 should be worked as "sight" exercises, $\frac{1}{4}$ being used for 25%, $\frac{1}{8}$ for $12\frac{1}{2}$%, etc.

810. First compute the interest for one year.
    1. $3.60 for a year; $\frac{1}{6}$ of $3.60 for 2 months.
    2. $3.60 for a year; $\frac{1}{6}$ of $3.60 for 60 days.
    3. $5.00 for a year; $5 × $2\frac{1}{2}$ for 2 yr. 6 mo.
    4. $6.00 for a year; $\frac{1}{12}$ of $6 for 30 days.
    5. $9.00 for a year; $\frac{1}{4}$ of $9 for 90 days.

812. To take advantage of any opportunities for cancellation that may be offered, this method is given. It will afterwards be found useful in calculating the principal, the rate, or the time.

Pupils should not at this stage be taught more than one method of finding interest, and that the most direct and the most obvious one.

813. 2. In changing 2 mo. 12 da. to the fraction of a year, it is not necessary to reduce to the lowest terms. Change the time to 72 days, and write 360 underneath, $\frac{72}{360}$; the necessary reduction can be made later in the cancellation. 6. Write 21 months as $\frac{21}{12}$ years, canceling afterwards.

The 100 in the denominator of an interest example should seldom be canceled, except as a whole or by 10.

815. 1. 6 hr. 17 min. 5 sec. = 22625 sec.; 3 hr. 15 min. 25 sec. = 11725 sec. *Ans.* $\frac{11725}{22625} = \frac{469}{905}$.

2. 3 mi. 96 rd. × $3\frac{1}{3}$ = 9 mi. 288 rd. + 1 mi. 32 rd. = 11 mi. *Ans.*

4. A furnished $\frac{1}{2}$ of the money, and should receive $\frac{1}{2}$ of $1500, or $750; B should receive $\frac{1}{3}$ of $1500, or $500; C should receive $\frac{1}{6}$ of $1500.

5. If 5 T. 1000 lb., or 11000 lb., cost $30.25, 1 lb. will cost $30.25 ÷ 11000; and 7 T. 320 lb., or 14320 lb., will cost ($30.25 ÷ 11000) × 14320. Cancel. $\dfrac{\$30.25 \times 14320}{11000}$

6. 25¢ × $8\frac{16}{25}$ = 25¢ × $8\frac{4}{5}$. *Ans.*

7. 2 yr. 7 mo. 8 da. = 31 mo. 8 da. = $31\frac{8}{30}$ mo. = $31\frac{4}{15}$ mo. $45 × $31\frac{4}{15}$ = *Ans.*

10. 360 yd. @ 30¢ cost $108. The number of square yards = 360 × $\frac{27}{36}$ = 360 × $\frac{3}{4}$ = 270, on which the duty at 8¢ per sq. yd. will be 8¢ × 270, or $21.60. The duty on the value will be 50% of $108, or $\frac{1}{2}$ of $108, or $54; the total duty being $21.60 + $54 = $75.60. *Ans.*

816. 3. 5 bbl., 300 lb. each, @ 5¢ per lb.
4. Interest on $200 for 6 mo. @ 6%.
5. 12 men take 24 days; how long will 24 men take?
6. What decimal of 640 acres is 320 acres?
7. 20 thousand bricks @ $20 per M.
8. 5600 lb. @ 87¢ per bu. of 56 lb.

NOTES ON CHAPTER NINE

    9. 10 lb. cost $8; find cost of 21 lb.
   10. Freight on 20 hundred lb. @ 70¢ per cwt.

817. 2. The wall 8 yd. × 4 yd. contains 32 sq. yd.; the door is $\frac{8}{3}$ yd. by $1\frac{1}{2}$ yd., and contains 4 sq. yd.; 32 sq. yd. − 4 sq. yd. = 28 sq. yd. *Ans.*

   3. Number of square inches in the surface of the widest face = 8 × 4; in the surface of one side = 8 × 2; in the surface of end = 4 × 2.

   4. $(288 \times 96) \div (8 \times 4)$.

   5. $[(24 \times 12) \times (8 \times 12)] \div [8 \times 2]$.

   6. See No. 20. Make four rectangles adjoining each other, each 8 inches high — the first and the third being 4 inches wide; and the second and the fourth, 2 inches wide. Above and below the second, and connected with it, draw rectangles 2 inches wide and 4 inches high. These two rectangles may be drawn above and below either of the other rectangles, the above dimension being used if drawn above and below the fourth; if drawn above and below the first or the third, they will be 4 inches wide and 2 inches high. The pupils should be permitted to make the diagram in their own way, and they should be encouraged to make one that differs from one drawn by a desk-mate.

   8. The number of rolls will be $(45 \times 36) \div (24 \times 1\frac{1}{2})$. Cancel.

818. The scholars should make this table without any assistance. To obtain the number of acres in a square mile, indicate the number of square rods in a square mile, 320 × 320, and divide by the number of square rods in an acre, 160.

   14. The number of yards = $(5+3+4+7+3+6+12+10) \times 5\frac{1}{2}$.

   15. Original dimensions 12 rods × 13 rods, making area 156 sq. rd. Present area = 156 sq. rd. − (15 + 21) sq. rd.

   19. $[\frac{1}{2} \text{ of } (80 \times 60\frac{1}{2})] \div 4840 = Ans.$

819. 1. 43 yd. = $(43 \div 5\frac{1}{2})$ rd. = $(43 \div \frac{11}{2})$ rd. = $(43 \times \frac{2}{11})$ rd. = $\frac{86}{11}$ rd. = $7\frac{9}{11}$ rd.

2. 43 yd. = $7\frac{9}{11}$ rd. $\frac{9}{11}$ rd. = $(\frac{9}{11} \times 5\frac{1}{2})$ yd. = $(\frac{9}{11} \times \frac{11}{2})$ yd. = $4\frac{1}{2}$ yd. *Ans.* 7 rd. $4\frac{1}{2}$ yd.

3. 43 yd. = 7 rd. $4\frac{1}{2}$ yd. = 7 rd. 4 yd. $1\frac{1}{2}$ ft.

4. 43 yd. = 7 rd. 4 yd. $1\frac{1}{2}$ ft. = 7 rd. 4 yd. 1 ft. 6 in.

824. 34. Carrying 1 to the column of yards, the total becomes 8 yd. or 1 rd. $2\frac{1}{2}$ yd. Changing $\frac{1}{2}$ yd. to 1 ft. 6 in., and adding this to 17 rd. 2 yd. 1 ft. 6 in., the accompanying answer is obtained.

    4 rd. 3 yd. 1 ft.
    9 rd. 4 yd. 2 ft.
    3 rd.   1 ft. 6 in.
   17 rd. $2\frac{1}{2}$ yd. 1 ft. 6 in.

38. 8 rd. 0 yd. 1 ft. $-$ 2 rd. 0 yd. 2 ft. = 5 rd. $4\frac{1}{2}$ yd. 2 ft. = 5 rd. 4 yd. 2 ft. $+$ 1 ft. 6 in. = 5 rd. 5 yd. 6 in. *Ans.*

   17 rd. 2 yd. 1 ft. 6 in.
    $+\frac{1}{2}$ yd. = 1 ft. 6 in.
   17 rd. 3 yd. *Ans.*

40. 5 rd. 4 yd. 2 ft. $\times$ 4 = 23 rd. $1\frac{1}{2}$ yd. 2 ft. = 23 rd. 1 yd. 2 ft. + 1 ft. 6 in. = 23 rd. 2 yd. 6 in.

825. 10. The other dimensions would be 8 ft. and 4 ft., or 16 ft. and 2 ft.

14. Number of cubic yards = $\frac{18}{3} \times \frac{55}{3} \times \frac{6}{3}$. Cancel.

15. $\frac{1}{2}$ (yd.) $\times$ 2 (yd.) $\times$ width (yd.) = 1 (cu. yd.); or $\frac{1}{2} \times 2 \times x = 1$; $x = 1$; 1 yd. *Ans.*

16. A gallon contains 231 cu. in.; a cubic foot contains 1728 cu. in. 1 cu. ft. = (1728 $\div$ 231) gal. = $7\frac{51}{231}$ gal. = about $7\frac{1}{2}$ gal. *Ans.*

17. About $1\frac{1}{4}$ cu. ft. *Ans.*

18. Number of gallons (21 $\times$ 15 $\times$ 22) $\div$ 231.

19. The decimal in the denominator is removed one place to the right, and a cipher is annexed to 64 in the numerator.

$$\frac{36 \times 28 \times 64 \odot 0}{2150 \odot 4}$$

NOTE. — 2150.4 cu. in. is used instead of 2150.42 cu. in., the more correct equivalent, because the former is divisible by 6, 7, 8, etc.

25. [$ 6.40 $\times$ (40 $\times$ $16\frac{1}{2}$) $\times$ 4 $\times$ 3] $\div$ $24\frac{3}{4}$. Cancel.

NOTES ON CHAPTER NINE 83

826. 3. At 7½ gal. to cu. ft., a tank of 150 gal. will contain (150 ÷ 7½) cu. ft. = 20 cu. ft. The dimensions will be 2 ft. × 2 ft. × 5 ft., or 4 ft. × 1 ft. × 5 ft., etc., etc.

4. At 1¼ cu. ft. to a bushel, the bin will contain 1¼ cu. ft. × 100 = 125 cu. ft. The dimensions will be 5 ft. × 5 ft. × 5 ft., or 5 ft. by 10 ft. by 2½ ft., etc., etc.

5. 1000 bricks will build (1000 ÷ 20) cu. ft. A wall 1 ft. thick can be 10 ft. long and 2 ft. high, or 5 ft. long and 4 ft. high, etc.

6. 10 yd. × 5 yd. × 2 yd. (30 ft. × 15 ft. × 6 ft.), 4 yd. × 5 yd. × 5 yd., etc.

7. A gallon weighs about 8 lb.; a pint about 1 lb.

8. A cubic foot of iron weighs about 7 times 64 pounds.

9. About ½ of $800.

10. About 4 years' interest.

831. 5. See Arithmetic, Art. 642.

6. See Arts. 829–830.

7. Three inches square = (3 × 3) sq. in.

832. 1. See Art. 1022, No. 15.

4. Including Sept. 19, the time is (28 + 30 + 31 + 30 + 31 + 31 + 19) days.

833. 6. The written analysis of an arithmetic example should be required occasionally as an exercise in composition.

7. 7000 gr. × 2⅞ = number of grains in 2 lb. 14 oz. Dividing by 480 grains, the number of Troy ounces is obtained — (7000 × 2⅞) ÷ 480. Multiplying $1.80 by this number, the cost of the urn is ascertained — ($1.80 × 7000 × 2⅞) ÷ 480.

835. 8. Since the denominator of a fraction indicates the number of parts into which a thing is divided, a larger denominator indicates a greater number of parts, and, therefore, smaller ones.

839. 1. If three-fifths of a bbl. cost $2.13, six-fifths will cost twice $2.13.

8. Commission at 1% would amount to $3; at $\frac{1}{2}$%, it is $1.50.

18. $\frac{3}{4} = 18$¢; $\frac{1}{4} = 6$¢; $\frac{1}{8}$, or $\frac{1}{2}$ of $\frac{1}{4}$, = $\frac{1}{2}$ of 6¢.

841. Add without re-writing the fractions reduced to a common denominator.

845. 10. 75% of $\frac{6}{10}$, or $\frac{3}{4}$ of $\frac{6}{10}$, or $\frac{9}{20}$, is sold for $1710. Factory is worth $1710 ÷ $\frac{9}{20}$.

13. First piece contains $(20 \times \frac{3}{4})$ sq. yd., or 15 sq. yd. Width of second piece in yards = 15 ÷ 12.

16. 8 men and 5 boys = 8 men + $2\frac{1}{2}$ men = $10\frac{1}{2}$ men. If 7 men do a piece of work in $10\frac{1}{2}$ days, 1 man will do it in $10\frac{1}{2}$ days × 7, and $10\frac{1}{2}$ men will do it in $(10\frac{1}{2}$ days × 7) ÷ $10\frac{1}{2}$. Cancel.

17. Each of the six square faces of a cube contains (6×6) sq. in., or 36 sq. in.; the whole surface will be, therefore, 36 sq. in. × 6. Each face contains $(\frac{1}{2} \times \frac{1}{2})$ sq. ft. = $\frac{1}{4}$ sq. ft.; whole surface = $\frac{1}{4}$ sq. ft. × 6.

Contents in cu. in. = $6 \times 6 \times 6$; in cu. ft. = $\frac{1}{2} \times \frac{1}{2} \times \frac{1}{2}$.

18. $\frac{27}{45}$ in water and $\frac{10}{45}$ in mud = $\frac{37}{45}$, leaving $\frac{8}{45}$ above water, or 5 ft. Length of post = 5 ft. + $\frac{8}{45}$.

19. $[(10 \times 9) + (12 \times 10) + (8 \times 11) + (6 \times 12) + (2 \times 13) + (1 \times 14)] ÷ [10 + 12 + 8 + 6 + 2 + 1]$.

21. From Oct. 25 to Dec. 31, inclusive, there are $7 + 30 + 31$, or 68 days; Oct. 30 is Sunday; also Nov. 6, 13, 20, 27; Dec. 4, 11, 18, and 25 — 9 Sundays, Election Day, and Thanksgiving to be deducted, or 11 days, leaving 57 days, at 3\frac{1}{2}$ per day.

23. 12 lb. tea cost $2.80 + $2.00, or $4.80; value per lb. 40¢.

24. House and lot, or $3\frac{1}{2}$ lots $+1$ lot, or $4\frac{1}{2}$ lots $= \$8100$; 1 lot $= \$8100 \div 4\frac{1}{2} = \$1800$; house $= \$1800 \times 3\frac{1}{2}$.

25. $\dfrac{\$18 \times (20 \times 12) \times (15 \times 12) \times (6 \times 12)}{1000 \times 8 \times 4 \times 2}$

27. 36 yd. 8 in. $= 1304$ in.; 13 yd. 1 ft. 9 in. $= 489$ in.; quantity left $= 1304$ in. $- 489$ in. $= 815$ in.; fraction left $= \frac{815}{1304}$ $= \frac{5}{8}$; decimal left $= .625$; per cent left $= 62\frac{1}{2}$. *Ans.*

28. Assessed value $= 80\%$ of $\$30000 = \$24000$. Taxes on 24 thousand dollars $= \$21.60 \times 24$.

846. 2. See Arithmetic, Art. 1251; angles *E*, *F*, *G*, and *H*; and *M*, *N*, *O*, and *P*.

3. Art. 1251; angles *A* and *B*, *C* and *D*.

4. Angles *I* and *J*, *K* and *L*. The scholars should understand that two lines can be perpendicular without one being a horizontal line and the other a vertical line.

5. The size of an angle does not depend upon the length of the lines that form the angle. Two short lines may meet at a very obtuse angle, and two long lines may form a very acute angle.

13. If the pupils have in their drawing lessons constructed triangles by means of compasses, these may be used; otherwise, let them manage as best they can, no great accuracy being required.

15. Children are accustomed to seeing an isosceles triangle in only one position: they should learn that if a triangle has two equal sides, it is isosceles, no matter whether the unequal side is vertical, horizontal, or oblique.

16–22. Accustom the scholars to the occasional employment of an oblique line as a base in constructing squares, rectangles, etc. See Arithmetic, Art. 1265. A card may be used to make a square corner.

24. See Arithmetic, Art. 929, No. 8, for a rectangle, a rhombus, and a rhomboid, having equal bases and equal altitudes. No. 5 shows three rhomboids of equal bases and equal altitudes, but differing in shape.

25. See Art. 929, No. 8.

847. 1. ($\frac{1}{2}$ of 15 × 20) sq. in. The length of the third side does not enter into the computation.

6. Let the scholars find the area of the rectangle, 66 ft. by 63 ft., and the two triangles, 31 ft. each by 63 ft., and find the sum of the areas. Then lead them to see that bringing the right-hand triangle to the left of the rhombus would make a rectangle 97 ft. by 63 ft., whose area is the sum above found.

7. Find the area in square meters, saying nothing more about the meter than that it is largely used on the continent of Europe, and is a little longer than a yard.

8-10. Give no rules yet for calculating the areas of trapezoids and trapeziums. Let the pupils ascertain the areas of the figures from the data supplied.

XIII

NOTES ON CHAPTER TEN

The formal study of algebra belongs to the high-school; but some so-called arithmetical problems are so much simplified by the use of the equation that it is a mistake for a teacher not to avail herself of this means of lightening her pupils' burdens.

In beginning this part of her mathematical instruction, the teacher should not bewilder her scholars with definitions. The necessary terms should be employed as occasion requires, and without any explanation beyond that which is absolutely necessary.

849. Very young pupils can give answers to most of these questions; so that there will be no need, for the present, at least, of introducing a number of axioms to enable the scholar to obtain a result that he can reach without them.

850. Pupils will learn how to work these problems by working a number of them. They may need to be told that x stands for $1x$; and that, as a rule, only abstract numbers are used in the equations, the denomination — dollars, marbles, etc. — being supplied afterwards.

While the scholars should be required to furnish rather full solutions of the earlier problems, they should be permitted to shorten the work by degrees, writing only whatever may be necessary.

4. $x + 2x = 54.$
5. $x + 5x = 78.$
6. $7x + 5x = 156.$
7. $9x - 3x = 66.$
8. $x + 2x + 6x = 27000.$
9. $x + 5x = 72.$
10. $x + 2x + 3x = 54.$
11. $x + 6x = 42.$

12. $2x + 10x = 96$.

13. Let $x =$ the fourth; then $4x =$ the third, $12x =$ the second, and $24x =$ the first.
$$x + 4x + 12x + 24x = 41.$$

14. $x =$ the second, $2x =$ first, $9x =$ third.

15. $5x + 4x = 81$.   17. $4x = 340$.

16. $24x = 456$.   19. $3x + 4x = 175$.

20. Let $x =$ each boy's share; $2x =$ each girl's share.
$$2x + 4x = 240.$$

21. $x =$ number of days son worked; $2x =$ number father worked. $3x =$ son's earnings; $8x =$ father's earnings.
$$3x + 8x = 165.$$

22. $x =$ number of dimes; $2x =$ number of nickels; $6x =$ number of cents.
$$(10 \times x) + (5 \times 2x) + (1 \times 6x) = 78,$$
or $\qquad\qquad 10x + 10x + 6x = 78.$

23. $15x - 12x = 75$.

24. $x + 4x + x + 4x = 250$.

25. Let $x =$ cost of speller; then $3x =$ cost of reader.

26. Let $x =$ smaller; then $5x =$ larger.

27. Let $x =$ Susan's number; $2x =$ Mary's; $3x =$ Jane's.

851. $10 : \tfrac{1}{3}x$ is the same as $\dfrac{x}{3}$.

852. Pupils already know that $\tfrac{3}{4}$ means $3 \div 4$, so that they can understand that $\dfrac{3x}{4}$ means $3x \div 4$, or $\tfrac{1}{4}$ of $3x$. When $\tfrac{1}{4}$ of something $(3x)$ is 24, the whole thing $(3x)$ must be 4 times 24, or 96; that is, when $\dfrac{3x}{4} = 24$, $3x = 96$.

When $\qquad \dfrac{2y}{3} = 24$, $2y = 24 \times 3$, or 72.

When $\qquad \dfrac{4z}{5} = 20$, $4z = 20 \times 5$, or 100.

From these examples can be formulated the rule for disposing of a fraction in one term of an equation, which is, to multiply both terms by the denominator of the fraction. In changing the first term of the equation, $\frac{3x}{4} = 24$, to $3x$, it has been multiplied by 4, so that the second term must also be multiplied by 4.

853. In solving these examples by the algebraic method of "clearing of fractions," attention may be called to its similarity to the arithmetical method. To find the value of y in **2**, the pupil multiplies 8 by 5 and divides the product by 2; as an example in arithmetic, he would divide 8 by $\frac{2}{5}$, that is, he would multiply 8 by $\frac{5}{2}$; the only difference being that by the latter method he would cancel.

While $\frac{2y}{5} = 8$ may be changed to $\frac{y}{5} = 4$ by dividing both terms by 2, beginners are usually advised to begin by "clearing of fractions," short methods being deferred to a later stage.

854. **6** may be written $\frac{3x}{5} + \frac{5x}{7} = 92$.

8. $2\frac{7}{8}x$ should be reduced to an improper fraction, making the equation, $\frac{23x}{8} = 115$. Make similar changes in **12, 14, 18,** and **20**.

855. 2. $x + \frac{5x}{2} = 100$.

5. $\frac{x}{2} + \frac{x}{4} = \frac{267}{4}$; $2x + x = 267$.

6. $\frac{3x}{4} - \frac{3x}{5} = 15$.

9. Let $5x$ = numerator; $7x$ = denominator. $7x - 5x = 24$; $2x = 24$; $x = 12$. The numerator, $5x$, will be 5 times 12, or 60; the denominator will be 84; and the fraction, $\frac{60}{84}$. *Ans.*

10. Let x = greater; $\frac{x}{7}$ = less.

$$x + \frac{x}{7} = 480.$$

Clearing of fractions, $\quad 7x + x = 3360,$

$$8x = 3360,$$
$$x = 420, \text{ the greater number,}$$
$$\frac{x}{7} = 60, \text{ the less.}$$

Or, $\qquad$ let $x =$ less; $7x =$ greater.
$$x + 7x = 480,$$
$$8x = 480,$$
$$x = 60, \text{ the less,}$$
$$7x = 420, \text{ the greater.}$$

The employment of the latter plan does away with fractions in the original equation.

11. $30x - x = 522,$ or $x - \dfrac{x}{30} = 522.$

13. Let $x =$ number of plums; $4x =$ number of peaches. Then $2x$ will be cost of plums, and $12x$ the cost of the peaches.
$$2x + 12x = 70.$$

15. $\qquad\qquad\qquad x - \dfrac{3x}{7} = 80.$

17. $\qquad\qquad x - \dfrac{3x}{8} - \dfrac{x}{4} = 24.$

18. $\qquad\qquad x + 1\tfrac{1}{2}x + (1\tfrac{1}{2}x \times 3\tfrac{1}{3}) = 15.$

$\qquad\qquad x + \dfrac{3x}{2} + 5x = 15.$

19. Let $x =$ price per yard of the 48-yard piece; $2x =$ price per yard of the 36-yard piece; $48x$ will be the total cost of one, and $72x$, of the other.
$$48x + 72x = 240.$$

20. $160x + 120x = 840.$

856. The pupils should be permitted to give these answers without assistance.

In Art. 857 is explained what is meant by "transposing."

NOTES ON CHAPTER TEN

858. While these exercises are so simple that they can be worked without a pencil, they should be used to show the steps generally taken in more complicated equations. In **1**, for instance, the work should take the form here indicated, only a single step being taken at a time.

$$x + 37 = 56$$
$$x = 56 - 37$$
$$x = 19$$

In **19**, the first step is to clear the equation of fractions by multiplying by 6; the second step is to transpose the unknown quantities to the left side of the equation, and the known quantities to the right; the third step is to combine the unknown quantities into one, and to make a similar combination of the known quantities; the last step is to find the value of x.

$$2x - 6 = 16 + \frac{x}{2} - \frac{x}{3}$$
$$12x - 36 = 96 + 3x - 2x$$
$$12x - 3x + 2x = 96 + 36$$
$$11x = 132$$
$$x = 12$$

After a little more familiarity with exercises of this kind, the pupil can take short cuts with less danger of mistakes; for the present, however, it will be safer to proceed in the slower way.

859. **5.** $x + (x + 75) + x + (x + 75) = 250.$
$$x + x + x + x = 250 - 75 - 75.$$

NOTE.—The parentheses used here are unnecessary. They are employed merely to show that $x + 75$ is one side of the field.

6. $x + (x + 8) = 86.$ **9.** $x + x + 72 = 96.$

7. $x + x + 318 = 2436.$ **10.** $x - \frac{x}{3} - \frac{x}{4} = 45.$

8. $x + \frac{x}{2} + 7 = 100.$

11. $x =$ one part; $2x - 6 =$ other part.
$$x + 2x - 6 = 45.$$

12. $x =$ John's money; $x + 5 =$ William's money.
$$3x + 15 + 5x = 103.$$

13. Let $x =$ price of a horse; $\quad x - 80 =$ price of a cow;
$4x =$ cost of four horses; $\quad 3x - 240 =$ cost of three cows.
$$4x + 3x - 240 = 635,$$
$$7x = 635 + 240 = 875,$$
$$x = 125, \text{ price, in dollars, of a horse};$$
$$x - 80 = 45, \text{ price, in dollars, of a cow}.$$

Other pupils may solve the problems in this way:
$x =$ price of a cow; $x + 80 =$ price of a horse.
$$3x + 4x + 320 = 635,$$
$$7x = 635 - 320 = 315,$$
$$x = 45, \text{ price, in dollars, of a cow};$$
$$x + 80 = 125, \text{ price, in dollars, of a horse}.$$

14. $x =$ number of dimes; $x + 11 =$ number of five-cent pieces; $10x =$ value of dimes (in cents); $5x + 55 =$ value of five-cent pieces.
$$10x + 5x + 55 = 100.$$

15. $x =$ greater; $x - 48 =$ less.
$$x + x - 48 = 100.$$
Or, $x =$ less; $x + 48 =$ greater.
$$x + x + 48 = 100.$$

17. $x =$ share of the first;
$x + 2400 =$ share of the second;
$x + 2400 + 2400 =$ share of the third.
$$x + x + 2400 + x + 2400 + 2400 = 18000.$$

18. Let $x =$ less; $x + 33 =$ greater.
$$x + 33 - 3x = 11.$$
Bringing known quantities to the left side of the equation, and the unknown quantities to the right,
$$33 - 11 = 3x - x,$$
$$22 = 2x,$$
$$11 = x.$$

Or, $\qquad x - 3x = 11 - 33,$
$$-2x = -22.$$
Changing signs of both terms,
$$2x = 22,$$
$$x = 11.$$
This problem may also be worked in this way:
$$x = \text{less};\ 3x + 11 = \text{greater}.$$
$$3x + 11 - x = 33.$$

19. $x =$ number of 5-cent stamps; $x + 15 =$ number of 2-cent stamps; $x + 30 =$ number of postal cards.
$$5x + 2x + 30 + x + 30 = 100.$$

20. $x =$ number of horses; $x + 17 =$ number of cows; $2x + 39 =$ number of sheep.
$$x + x + 17 + 2x + 39 = 88.$$

SUPPLEMENT

DEFINITIONS, PRINCIPLES, AND RULES

A **Unit** is a single thing.
A **Number** is a unit or a collection of units.
The **Unit of a Number** is one of that number.
Like Numbers are those that express units of the same kind.
Unlike Numbers are those that express units of different kinds.
A **Concrete Number** is one in which the unit is named.
An **Abstract Number** is one in which the unit is not named.
Notation is expressing numbers by characters.
Arabic Notation is expressing numbers by figures.
Roman Notation is expressing numbers by letters.
Numeration is reading numbers expressed by characters.
The **Place of a Figure** is its position in a number.

A figure standing alone, or in the first place at the right of other figures, expresses *ones*, or *units of the first order*.

A figure in the second place expresses *tens*, or *units of the second order*.

A figure in the third place expresses *hundreds*, or *units of the third order;* and so on.

A **Period** is a group of three orders of units, counting from right to left.

RULE FOR NOTATION. — *Begin at the left, and write the hundreds, tens, and units of each period in succession, filling vacant places and periods with ciphers.*

Rule for Numeration. — *Beginning at the right, separate the number into periods.*

Beginning at the left, read the numbers in each period, giving the name of each period except the last.

ADDITION

Addition is finding a number equal to two or more given numbers.

Addends are the numbers added.

The **Sum,** or **Amount,** is the number obtained by addition.

Principle. — *Only like numbers, and units of the same order can be added.*

Rule. — *Write the numbers so that units of the same order shall be in the same column.*

Beginning at the right, add each column separately, and write the sum, if less than ten, under the column added.

When the sum of any column exceeds nine, write the units only, and add the ten or tens to the next column.

Write the entire sum of the last column.

SUBTRACTION

Subtraction is finding the difference between two numbers.

The **Subtrahend** is the number subtracted.

The **Minuend** is the number from which the subtrahend is taken.

The **Remainder,** or **Difference,** is the number left after subtracting one number from another.

Principles. — *Only like numbers and units of the same order can be subtracted.*

The sum of the difference and the subtrahend must equal the minuend.

Rules. — I. *Write the subtrahend under the minuend, placing units of the same order in the same column.*

Beginning at the right, find the number that must be added to the first figure of the subtrahend to produce the figure in the corresponding order of the minuend, and write it below. Proceed in this way until the difference is found.

If any figure in the subtrahend is greater than the corresponding figure in the minuend, find the number that must be added to the former to produce the latter increased by ten; then add one to the next order of the subtrahend and proceed as before.

II. *Beginning at the units' column, subtract each figure of the subtrahend from the corresponding figure of the minuend and write the remainder below.*

If any figure of the subtrahend is greater than the corresponding figure in the minuend, add ten to the latter and subtract; then, (a) add one to the next order of the subtrahend and proceed as before; or, (b) subtract one from the next order of the minuend and proceed as before.

MULTIPLICATION

Multiplication is taking one number as many times as there are units in another number.

The **Multiplicand** is the number taken or multiplied.

The **Multiplier** is the number that shows how many times the multiplicand is taken.

The **Product** is the result obtained by multiplication.

PRINCIPLES. — *The multiplier must be an abstract number.*

The multiplicand and the product are like numbers.

The product is the same in whatever order the numbers are multiplied.

RULE. — *Write the multiplier under the multiplicand, placing units of the same order in the same column.*

Beginning at the right, multiply the multiplicand by the number of units in each order of the multiplier in succession. Write the figure of the lowest order in each partial product under the figure of the multiplier that produces it. Add the partial products.

To multiply by 10, 100, 1000, etc.

RULE. — *Annex as many ciphers to the multiplicand as there are ciphers in the multiplier.*

DIVISION

Division is finding how many times one number is contained in another, or finding one of the equal parts of a number.

The **Dividend** is the number divided.

The **Divisor** is the number contained in the dividend.

The **Quotient** is the result obtained by division.

PRINCIPLES. — *When the divisor and the dividend are like numbers, the quotient is an abstract number.*

When the divisor is an abstract number, the dividend and the quotient are like numbers.

The product of the divisor and the quotient, plus the remainder, if any, is equal to the dividend.

RULE. — *Write the divisor at the left of the dividend with a line between them.*

Find how many times the divisor is contained in the fewest figures on the left of the dividend, and write the result over the last figure of the partial dividend. Multiply the divisor by this quotient figure, and write the product under the figures divided. Subtract the product from the partial dividend used, and to the remainder annex the next figure of the dividend for a new dividend.

Divide as before until all the figures of the dividend have been used.

If any partial dividend will not contain the divisor, write a cipher in the quotient, and annex the next figure of the dividend.

If there is a remainder after the last division, write it after the quotient with the divisor underneath.

FACTORING

An **Exact Divisor** of a number is a number that will divide it without a remainder.

An **Odd Number** is one that cannot be exactly divided by two.

An **Even Number** is one that can be exactly divided by two.

The **Factors** of a number are the numbers that multiplied together produce that number.

A **Prime Number** is a number that has no factors.

A **Composite Number** is a number that has factors.

A **Prime Factor** is a prime number used as a factor.

A **Composite Factor** is a composite number used as a factor.

Factoring is separating a number into its factors.

To find the Prime Factors of a Number.

RULE. — *Divide the number by any prime factor. Divide the quotient, if composite, in like manner; and so continue until a prime quotient is found. The several divisors and the last quotient will be the prime factors.*

CANCELLATION

Cancellation is rejecting equal factors from dividend and divisor.

PRINCIPLE. — *Dividing dividend and divisor by the same number does not affect the quotient.*

GREATEST COMMON DIVISOR

A **Common Factor** (divisor or measure) is a number that is a factor of each of two or more numbers.

A **Common Prime Factor** is a prime number that is a factor of each of two or more numbers.

The **Greatest Common Factor** (divisor or measure) is the largest number that is a factor of each of two or more numbers.

Numbers are **prime to each other** when they have no common factor.

The greatest common divisor of two or more numbers is the product of their common prime factors.

PRINCIPLES. — *A common divisor of two numbers is a divisor of their sum, and also of their difference.*

A divisor of a number is a divisor of every multiple of that number; and a common divisor of two or more numbers is a divisor of any of their multiples.

To find the Common Prime Factors of Two or More Numbers.

RULE. — *Divide the numbers by any common prime factors, and the quotients in like manner, until they have no common factor; the several divisors are the common prime factors.*

To find the Greatest Common Divisor of Numbers that are Easily Factored.

RULE. — *Separate the numbers into their prime factors; the product of those that are common is the greatest common divisor.*

To find the Greatest Common Divisor of Numbers that are not Easily Factored.

RULE. — *Divide the greater number by the less; then divide the last divisor by the last remainder, continuing until there is no remainder. The last divisor is the greatest common divisor.*

If there are more than two numbers, find the greatest common divisor of two of them; then of that divisor and another of the numbers until all of the numbers have been used. The last divisor is the greatest common divisor.

LEAST COMMON MULTIPLE

A **Multiple** of a number is a number that exactly contains that number.

A **Common Multiple** of two or more numbers is a number that is a multiple of each of them.

The **Least Common Multiple** of two or more numbers is the smallest number that is a common multiple of them.

PRINCIPLES. — *A multiple of a number contains all the prime factors of that number.*

A common multiple of two or more numbers contains each of the prime factors of those numbers.

The Least Common Multiple of two or more numbers contains only the prime factors of each of the numbers.

To find the Least Common Multiple of Two or More Numbers.

RULE. — *Divide by any prime number that is an exact divisor of two or more of the numbers, and write the quotients and undivided numbers below. Divide these numbers in like manner, continuing until no two of the remaining numbers have a common factor. The product of the divisors and remaining numbers is the least common multiple.*

FRACTIONS

A **Fraction** is one or more of the equal parts of anything.

The **Unit of a Fraction** is the number or thing that is divided into equal parts.

A **Fractional Unit** is one of the equal parts into which the number or thing is divided.

The **Terms of a Fraction** are its numerator and its denominator.

The **Denominator** of a fraction shows into how many parts the unit is divided.

The **Numerator** of a fraction shows how many of the parts are taken.

A fraction indicates division; the numerator being the dividend and the denominator the divisor.

The **Value of a Fraction** is the quotient of the numerator divided by the denominator.

Fractions are divided into two classes — **Common** and **Decimal.**

A **Common Fraction** is one in which the unit is divided into any number of equal parts.

A common fraction is expressed by writing the numerator above the denominator with a dividing line between.

Common fractions consist of three principal classes — **Simple, Compound,** and **Complex.**

A **Simple Fraction** is one whose terms are whole numbers.

A **Proper Fraction** is a simple fraction whose numerator is less than its denominator.

An **Improper Fraction** is a simple fraction whose numerator equals or exceeds its denominator.

A **Compound Fraction** is a fraction of a fraction.

A **Complex Fraction** is one having a fraction in its numerator, or in its denominator, or in both.

A **Mixed Number** is a whole number and a fraction written together.

The **Reciprocal of a Number** is one divided by that number.

The **Reciprocal of a Fraction** is one divided by the fraction, or the fraction inverted.

PRINCIPLES. — *Multiplying the numerator or dividing the denominator multiplies the fraction.*

Dividing the numerator or multiplying the denominator divides the fraction.

Multiplying or dividing both terms of a fraction by the same number does not alter the value of the fraction.

Reduction of fractions is changing their terms without altering their value.

To reduce a Fraction to Higher Terms.

RULE. — *Multiply both numerator and denominator by the same number.*

To reduce a Fraction to its Lowest Terms.

RULE. — *Divide both terms of the fraction by their greatest common divisor.*

A fraction is in its lowest terms when the numerator and the denominator are prime to each other.

To reduce a Mixed Number to an Improper Fraction.

RULE. — *Multiply the whole number by the denominator; to the product add the numerator; and place the sum over the denominator.*

To reduce an Improper Fraction to a Whole or to a Mixed Number.

RULE. — *Divide the numerator by the denominator.*

A **Common Denominator** is a denominator common to two or more fractions.

The **Least Common Denominator** is the smallest denominator common to two or more fractions.

To reduce Fractions to their Least Common Denominator.

RULE. — *Find the least common multiple of all the denominators for the least common denominator. Divide this multiple by the denominator of each fraction, and multiply the numerator by the quotient.*

ADDITION OF FRACTIONS

PRINCIPLE. — *Only like fractions can be added.*

RULE. — *Reduce the fractions, if necessary, to a common denominator, and over it write the sum of the numerators.*

If there are mixed numbers, add the fractions and the whole numbers separately, and unite the results.

SUBTRACTION OF FRACTIONS

PRINCIPLE. — *Only like fractions can be subtracted.*

RULE. — *Reduce the fractions, if necessary, to a common denominator, and over it write the difference between the numerators.*

If there are mixed numbers subtract the fractions and the whole numbers separately, and unite the results.

MULTIPLICATION OF FRACTIONS

RULE. — *Reduce whole and mixed numbers to improper fractions; cancel the factors common to numerators and denominators, and write the product of the remaining factors in the numerators over the product of the remaining factors in the denominators.*

DIVISION OF FRACTIONS

Rules. — I. *Reduce whole and mixed numbers to improper fractions. Reduce the fractions to a common denominator. Divide the numerator of the dividend by the numerator of the divisor.*

II. *Invert the divisor and proceed as in multiplication of fractions.*

To reduce a Complex Fraction to a Simple One.

Rules. — I. *Multiply the numerator of the complex fraction by its denominator inverted.*

II. *Multiply both terms by the least common multiple of the denominators.*

DECIMALS

A **Decimal Fraction** is one in which the unit is divided into tenths, hundredths, thousandths, etc.

A **Decimal** is a decimal fraction whose denomination is indicated by the number of places at the right of the decimal point.

The **Decimal Point** is the mark used to locate units.

A **Mixed Decimal** is a whole number and a decimal written together.

A **Complex Decimal** is a decimal with a common fraction written at its right.

To write Decimals.

Rule. — *Write the numerator; and from the right, point off as many decimal places as there are ciphers in the denominator, prefixing ciphers, if necessary, to make the required number.*

To read Decimals.

Rule. — *Read the numerator, and give the name of the right-hand order.*

Principles. — *Prefixing ciphers to a decimal diminishes its value.*

*Removing ciphers from the left of a decimal increases its value.
Annexing ciphers to a decimal or removing ciphers from its right does not alter its value.*

To reduce a Decimal to a Common Fraction.

RULE.— *Write the figures of the decimal for the numerator, and 1, with as many ciphers as there are places in the decimal, for the denominator, and reduce the fraction to its lowest terms.*

To reduce a Common Fraction to a Decimal.

RULE.— *Annex decimal ciphers to the numerator, and divide it by the denominator.*

To reduce Decimals to a Common Denominator.

RULE. — *Make their decimal places equal by annexing ciphers.*

ADDITION AND SUBTRACTION OF DECIMALS

Decimals are added and subtracted the same as whole numbers.

MULTIPLICATION OF DECIMALS

RULE. — *Multiply as in whole numbers, and from the right of the product, point off as many decimal places as there are decimal places in both factors.*

DIVISION OF DECIMALS

RULE.— *Make the divisor a whole number by removing the decimal point, and make a corresponding change in the dividend. Divide as in whole numbers, and place the decimal point in the quotient under (or over) the new decimal point in the dividend.*

ACCOUNTS AND BILLS

A **Debtor** is a person who owes another.
A **Creditor** is a person to whom a debt is due.

An **Account** is a record of debits and credits between persons doing business.

The **Balance** of an account is the difference between the debit and credit sides.

A **Bill** is a written statement of an account.

An **Invoice** is a written statement of items, sent with merchandise.

A **Receipt** is a written acknowledgment of the payment of part or all of a debt.

A bill is receipted when the words, "Received Payment," are written at the bottom, signed by the creditor, or by some person duly authorized.

DENOMINATE NUMBERS

A **Measure** is a standard established by law or custom, by which distance, capacity, surface, time, or weight is determined.

A **Denominate Unit** is a unit of measure.

A **Denominate Number** is a denominate unit or a collection of denominate units.

A **Simple Denominate Number** consists of denominate units of one kind.

A **Compound Denominate Number** consists of denominate units of two or more kinds.

A **Denominate Fraction** is a fraction of a denominate number.

A denominate fraction may be either **common** or **decimal.**

Reduction of denominate numbers is changing them from one denomination to another without altering their value.

Reduction Descending is changing a denominate number to one of a lower denomination.

RULE. — *Multiply the highest denomination by the number required to reduce it to the next lower denomination, and to the product add the units of that lower denomination, if any. Proceed in this manner until the required denomination is reached.*

Reduction Ascending is changing a denominate number to one of a higher denomination.

RULE. — *Divide the given denomination successively by the numbers that will reduce it to the required denomination. To this quotient annex the several remainders.*

To find the Time between Dates.

RULE. — *When the time is less than one year, find the exact number of days; if greater than one year, find the time by compound subtraction, taking* 30 *days to the month.*

PERCENTAGE

Per Cent means hundredths.

Percentage is computing by hundredths.

The elements involved in percentage are the **Base, Rate, Percentage, Amount,** and **Difference.**

The **Base** is the number of which a number of hundredths is taken.

The **Rate** indicates the number of hundredths to be taken.

The **Percentage** is one or more hundredths of the base.

The **Amount** is the base increased by the percentage.

The **Difference** is the base diminished by the percentage.

To find the Percentage when the Base and Rate are Given.

RULE. — *Multiply the base by the rate expressed as hundredths.*

To find the Rate when the Percentage and Base are Given.

RULE. — *Divide the percentage by the base.*

To find the Base when the Percentage and Rate are Given.

RULE. — *Divide the percentage by the rate expressed as hundredths.*

To find the Base when the Amount and Rate are Given.

RULE. — *Divide the amount by* 1 + *the rate expressed as hundredths.*

To find the Base when the Difference and Rate are Given.

RULE. — *Divide the difference by 1 — the rate expressed as hundredths.*

PROFIT AND LOSS

Profit or **Loss** is the difference between the buying and selling prices.

In Profit and Loss,

The buying price, or cost, is the *base*.
The rate per cent profit or loss is the *rate*.
The profit or loss is the *percentage*.
The selling price is the *amount* or *difference*, according as it is more or less than the buying price.

COMMERCIAL DISCOUNT

Commercial Discount is a percentage deducted from the list price of goods, the face of a bill, etc.

The **Net Price** of goods is the sum received for them.

In Commercial Discount,

The list price, or
The face of the bill } is the *base*.

The rate per cent discount is the *rate*.
The discount is the *percentage*.
The list price diminished by the discount is the *difference*.

In successive discounts, the first discount is made from the list price or the face of the bill; the second discount, from the list price or face of the bill diminished by the first discount; and so on.

COMMISSION

Commission is a percentage allowed an agent for his services.

A **Commission Agent** is one who transacts business on commission.

A **Consignment** is the merchandise forwarded to a commission agent.

The **Consignor** is the person who sends the merchandise.

The **Consignee** is the person to whom the merchandise is sent.

The **Net Proceeds** is the sum remaining after all charges have been deducted.

In buying, the commission is a percentage of the *buying price;* in selling, a percentage of the *selling price;* in collecting, a percentage of the *sum collected;* hence:

The sum invested, or
The sum collected } is the *base.*

The rate per cent commission is the *rate.*

The commission is the *percentage.*

The sum invested increased by the commission is the *amount.*

The sum collected diminished by the commission is the *difference.*

INSURANCE

Insurance is a contract of indemnity.

Insurance is of three kinds — **Fire, Marine,** and **Life.**

Fire Insurance is indemnity against loss of property by fire.

Marine Insurance is indemnity against loss of property by the casualities of navigation.

Life Insurance is indemnity against loss of life.

The **Insurance Policy** is the contract setting forth the liability of the insurer.

The **Policy Face** is the amount of insurance.

The **Premium** is the price paid for insurance.

The **Insurer,** or **Underwriter,** is the company issuing the policy.

The **Insured** is the person for whose benefit the policy is issued.

In Insurance,

The policy face is the *base.*

The rate per cent premium is the *rate.*

The premium is the *percentage.*

TAXES

A **Tax** is a sum of money levied on persons or property for public purposes.

A **Personal**, or **Poll Tax**, is a tax on the person.

A **Property Tax** is a tax of a certain per cent on the assessed value of property.

Property may be either personal or real.

Personal Property consists of such things as are movable.

Real Property is that which is fixed, or immovable.

In Taxes,

The assessed value is the *base*.
The rate of taxation is the *rate*.
The tax is the *percentage*.

DUTIES

Duties are taxes on imported goods.

Duties are either **Specific** or **Ad Valorem**.

A **Specific Duty** is a tax on goods without regard to cost.

An **Ad Valorem** duty is a tax of a certain per cent on the cost of goods.

In Ad Valorem Duties,

The cost of the goods is the *base*.
The rate per cent duty is the *rate*.
The ad valorem duty is the *percentage*.

INTEREST

Interest is the sum paid for the use of money.

The **Principal** is the sum loaned.

The **Amount** is the sum of the principal and interest.

The **Rate of Interest** is the rate per cent for one year.

The **Legal Rate** is the rate fixed by law.

Usury is interest at a higher rate than that fixed by law.

Simple Interest is interest on the principal only.

DEFINITIONS, PRINCIPLES, AND RULES

To find the Interest when the Principal, Time, and Rate are Given.

RULE. — *Multiply the principal by the rate expressed as hundredths, and this product by the time expressed in years.*

To find the Time when the Principal, Interest, and Rate are Given.

RULE. — *Divide the given interest by the interest for one year.*

To find the Rate when the Principal, Interest, and Time are Given.

RULE. — *Divide the given interest by the interest at one per cent.*

To find the Principal when the Interest, Rate, and Time are Given.

RULE. — *Divide the given interest by the interest on $1.*

To find the Principal when the Amount and Time and Rate are Given.

RULE. — *Divide the given amount by the amount of $1.*

INTEREST BY ALIQUOT PARTS.

To find the Interest for Years, Months, and Days.

RULE. — *Find the interest for one year and take this as many times as there are years.*

Take the greatest number of the given months that equals an aliquot part of a year and find the interest for this time. Take aliquot parts of this for the remaining months.

In the same manner find the interest for the days.

The sum of these interests will be the interest required.

To find the Interest when the Time is Less than a Year.

RULE. — *Find the interest for the time in months or days that will gain one per cent of the principal.*

Find by aliquot parts, as in the first rule, the interest for the remaining time.

The sum of these interests will be the interest required.

Interest by Six Per Cent Method.

To find the Interest at 6%.

RULE.— For Years: *Multiply the principal by the rate expressed as hundredths, and that product by the number of years.*

For Months: *Move the decimal point two places to the left, and multiply by one-half the number of months.*

For Days: *Move the decimal point three places to the left, and multiply by one-sixth the number of days.*

To find the interest at any other rate per cent, divide the interest at 6% by 6, and multiply the quotient by the given rate.

To find Exact Interest.

RULE.— *Multiply the principal by the rate expressed as hundredths, and that product by the time expressed in years of 365 days.*

ANNUAL INTEREST

Annual Interest is interest payable annually. If not paid when due, annual interest draws simple interest.

To find the Amount Due on a Note with Annual Interest, when the Interest has not been Paid Annually.

RULE.— *Find the interest on the principal for the entire time, and on each annual interest for the time it remained unpaid. The sum of the principal and all the interest is the amount due.*

COMPOUND INTEREST

Compound Interest is interest on the principal and on the unpaid interest, which is added to the principal at regular intervals. The interest may be compounded annually, semi-annually, quarterly, etc., according to agreement.

To find Compound Interest.

RULE.— *Find the amount of the given principal for the first period. Considering this as a new principal, find the amount of* it for the next period, continuing in this manner for the given time.

Find the difference between the last amount and the given principal, which will be the compound interest.

PARTIAL PAYMENTS

Partial Payments are part payments of a note or debt. Each payment is recorded on the back of the note or the written obligation.

UNITED STATES RULE. — *Find the amount of the principal to the time when the payment or the sum of two or more payments equals or exceeds the interest.*

From this amount deduct the payment or sum of payments.

Use the balance then due as a new principal, and proceed as before.

MERCHANTS' RULE. — *Find the amount of an interest-bearing note at the time of settlement.*

Find the amount of each credit from its time of payment to the time of settlement; subtract their sum from the amount of the principal.

BANK DISCOUNT

Bank Discount is a percentage retained by a bank for advancing money on a note before it is due.

The **Sum Discounted** is the face of the note, or if interest-bearing, the amount of the note at maturity.

The **Term of Discount** is the number of days from the day of discount to the day of maturity.

The **Bank Discount** is the interest on the sum discounted for the term of discount.

The **Proceeds** of a note is the sum discounted less the bank discount.

Problems in bank discount are calculated as problems in interest.

In Bank Discount,
The sum discounted is the *principal*.
The rate of discount is the *rate of interest*.
The term of discount is the *time*.
The bank discount is the *proceeds*.

EXCHANGE

Exchange is making payments at a distance by means of drafts or bills of exchange.

Domestic Exchange is exchange between places in the same country.

Foreign Exchange is exchange between different countries.

Exchange is *at par* when a draft, or bill, sells for its face value; *at a premium* when it sells for more than its face value; *at a discount* when it sells for less.

The cost of a sight draft is the face of the draft increased by the premium, or diminished by the discount.

The cost of a time draft is the face of the draft increased by the premium, or diminished by the discount, and this result diminished by the bank discount.

To find the Cost of a Draft.

RULE. — *Find the cost of $1 of the draft; multiply this by the face of the draft.*

To find the Face of a Draft.

RULE. — *Divide the cost of the draft by the cost of $1 of the draft.*

EQUATION OF PAYMENTS

Equation of Payments is a method of ascertaining at what time several debts due at different times may be settled by a single payment.

The **Equated Time** of payment is the time when the several debts may be equitably settled by one payment.

The **Term of Credit** is the time the debt has to run before it becomes due.

The **Average Term of Credit** is the time the debts due at different times have to run, before they may be equitably settled by one payment.

To find the **Equated Time of Payment** when the **Terms of Credit** begin at the **Same Date.**

RULE. — *Multiply each debt by its term of credit, and divide the sum of the products by the sum of the debts. The quotient will be the average term of credit.*

Add the average term of credit to the date of the debts, and the result will be the equated time of payment.

To find the **Equated Time** when the **Terms of Credit** begin at **Different Dates.**

RULE. — *Find the date at which each debt becomes due. Select the earliest date as a standard.*

Multiply each debt by the number of days between the standard date and the date when the debt becomes due, and divide the sum of the products by the sum of the debts. The quotient will be the average term of credit from the standard date.

Add the average term of credit to the standard date, and the result will be the equated time of payment.

RATIO

Ratio is the relation one number bears to another of the same kind.

The **Terms** of the ratio are the numbers compared.

The **Antecedent** is the first term.

The **Consequent** is the second term.

The antecedent and consequent form a *couplet*.

PRINCIPLES. — See Fractions.

PROPORTION

A **Proportion** is formed by two equal ratios.

The **Extremes** of a proportion are the first and last terms.

The **Means** of a proportion are the second and third terms.

PRINCIPLES. — *The product of the means is equal to the product of the extremes.*

Either mean equals the product of the extremes divided by the other mean.

Either extreme equals the product of the means divided by the other extreme.

RULE FOR PROPORTION. — *Represent the required term by x.*

Arrange the terms so that the required term and the similar known term may form one couplet, the remaining terms the other.

If the required term is in the extremes, divide the product of the means by the given extreme.

If the required term is in the means, divide the product of the extremes by the given mean.

PARTNERSHIP

Partnership is an association of two or more persons for business purposes.

The **Partners** are the persons associated.

The **Capital** is that which is invested in the business.

The **Assets** are the partnership property.

The **Liabilities** are the partnership debts.

To find the Profit, or Loss, of Each Partner when the Capital of Each is Employed for the Same Period of Time.

RULE. — *Find the part of the entire profit, or loss, that each partner's capital is of the entire capital.*

To find the Profit, or Loss, of Each Partner when the Capital of Each is Employed for Different Periods of Time.

RULE. — *Find each partner's capital for one month, by multiplying the amount he invests by the number of months it is employed; then find the part of the entire profit, or loss, that each partner's capital for one month is of the entire capital for one month.*

INVOLUTION

A **Power** of a number is the product obtained by using that number a certain number of times as a factor.

The **First Power** of a number is the number itself.

The **Second Power** of a number, or the **Square**, is the product of a number taken twice as a factor.

The **Third Power** of a number, or the **Cube**, is the product of a number taken three times as a factor.

An **Exponent** is a small figure written a little to the right of the upper part of a number to indicate the power.

Involution is finding any power of a number.

To find the Power of a Number.

RULE. — *Take the number as a factor as many times as there are units in the exponent.*

EVOLUTION

A **Root** is one of the equal factors of a number.

The **Square Root** of a number is one of its two equal factors.

The **Cube Root** of a number is one of its three equal factors.

Evolution is finding any root of a number.

Evolution may be indicated in two ways: by the *Radical Sign*, $\sqrt{\ }$, or by a *fractional exponent*.

The **Index** of a root is a small figure placed a little to the left of the upper part of the radical sign, to indicate what root is to be found. In expressing square root, the index is omitted.

In the fractional exponent, the numerator indicates the power to which the number is to be raised; the denominator indicates the root to be taken of the number thus raised.

To find the Square Root of a Number.

RULE. — *Point off in periods of two figures, commencing at units. Find the greatest square in the first period and place the root in the quotient. Subtract this square from the first period, and bring down the next period.*

Multiply the quotient figure by two, and use it as a trial divisor. Place the second figure in the quotient, and annex it also to the trial divisor. Then multiply the figures in the trial divisor by the second quotient figure, and subtract.

Bring down the next period, and proceed as before until the square root is found.

To find the Square Root of a Fraction.

RULE. — *Reduce the fraction to its simplest form, and find the square root of each term separately.*

To find the Cube Root of a Number.

RULE. — *Point off in periods of three figures each, beginning at units.*

Find the greatest cube in the first period and place the root in the quotient. Subtract this cube from the first period, and bring down the next period.

Multiply the square of the first quotient figure by three and annex two ciphers for a trial divisor. Place the second figure in the quotient. Then, to the trial divisor add three times the product of the first and second figures, also the square of the second. Multiply this sum by the second figure and subtract.

Bring down the next period, and proceed as before until the cube root is found.

To find the Cube Root of a Fraction.

RULE. — *Reduce the fraction to its simplest form, and find the cube root of each term separately.*

STOCKS AND BONDS.

Capital Stock is the money or property employed by a corporation in its business.

A **Share** is one of the equal divisions of capital stock.

The **Stockholders** are the owners of the capital stock.

The **Par Value** of stock is the face value.

The **Market Value** of stock is the sum for which it may be sold.

Stock is at a *premium* when the market value is above the par value; at a *discount*, when below par.

Bonds are interest-bearing notes issued by a government or a corporation.

A **Dividend** is a percentage apportioned among the stockholders.

A **Stock Broker** is a person who deals in stocks.

Brokerage is a percentage allowed a stock broker for his services.

In **Stocks and Bonds,**

    The par value is the *base*.

    The rate per cent premium, or discount, is the *rate*.

    The premium, discount, or dividend $\}$ is the *percentage*.

    The market value is the $\begin{cases} amount, \text{ or} \\ difference. \end{cases}$

NOTES, DRAFTS, AND CHECKS.

A **Promissory Note** is a written promise to pay a specified sum on demand, or at a specified time.

The **Face** of a note is the sum named in the note.

The **Maker** is the person who signs it.

The **Payee** is the person to whom the sum specified is to be paid.

The **Indorser** is the person who signs his name on the back of the note, thus becoming liable for its payment in case of default of the maker.

An **Interest-bearing Note** is one payable with interest.

If the words "with interest" are omitted, interest cannot be collected until after maturity.

A **Demand Note** is one payable when demand of payment is made.

A **Time Note** is one payable at a specified time.

A **Joint Note** is one signed by two or more persons who jointly promise to pay.

A **Joint and Several Note** is one signed by two or more persons who jointly and severally promise to pay.

In a joint note, each person is liable for the whole amount, but they must all be sued together. In the joint and several note, each is liable for the whole amount, and may be sued separately.

A **Negotiable Note** is one that may be transferred or sold. It contains the words " or bearer," or " or order."

A **Non-negotiable Note** is one not payable to the bearer, nor to the payee's order.

The **Maturity** of a note is the day on which it legally falls due.

A **Draft,** or **Bill of Exchange,** is a written order directing the payment of a specified sum of money.

The **Face** of a draft is the sum named in it.

The **Drawer** is the person who signs the draft.

The **Drawee** is the person ordered to pay the sum specified.

The **Payee** is the person to whom the sum specified is to be paid.

A **Sight Draft** is one payable when presented.

A **Time Draft** is one payable at a specified time.

An **Acceptance** of a time draft is an agreement by the drawee to pay the draft at maturity, which he signifies by writing across the face of the draft the word "accepted" with the date and his name.

A **Check** is an order on a bank or banker to pay a specified sum of money.

ANSWERS. — PART II.

Page 216.
16. $44\frac{5}{27}$.
17. $137\frac{3}{8}$.
18. $131\frac{5}{12}$.
19. $72\frac{1}{4}$.
20. $150\frac{1}{8}$.
21. $259\frac{3}{15}$.
22. $201\frac{1}{15}$.
23. $363\frac{1}{4}$.
24. $261\frac{7}{27}$.
25. $121\frac{5}{8}$.

Page 218.
1. $97\frac{9}{20}$.
2. $137\frac{1}{4}$.
3. $176\frac{3}{8}$.
4. $40\frac{1}{2}$.
5. $35\frac{7}{80}$.
6. $51\frac{11}{12}$.
7. $61\frac{5}{8}$.
8. $205\frac{3}{4}$.
9. 85.
10. $234\frac{1}{2}$.
11. $144\frac{1}{2}$.
12. $151\frac{1}{10}$.
13. $223\frac{1}{2}$.
14. $563\frac{7}{27}$.
15. $1096\frac{5}{8}$.
16. $749\frac{1}{2}$.
17. $1256\frac{11}{18}$.
18. $99\frac{5}{8}$.

19. $332\frac{7}{25}$.
20. 193.

Page 220.
11. $27\frac{7}{10}$.
12. $54\frac{3}{4}$.
13. $10\frac{1}{12}$.
14. $30\frac{1}{12}$.
15. $97\frac{1}{20}$.
16. $36\frac{1}{15}$.
17. $1\frac{8}{20}$.
18. $6\frac{5}{27}$.
19. $78\frac{7}{15}$.
20. $8\frac{11}{14}$.
21. $382\frac{1}{2}$.
22. $291\frac{3}{4}$.
23. $109\frac{3}{4}$.
24. $115\frac{3}{8}$.
25. $1\frac{1}{2}$.
26. $24\frac{4}{9}$.
27. $599\frac{5}{8}$.
28. $860\frac{7}{8}$.
29. $85\frac{7}{10}$.
30. $4\frac{10}{11}$.

Page 221.
31. $2\frac{3}{4}$.
32. $17\frac{1}{4}$.
33. $6\frac{3}{8}$.
34. $7\frac{5}{8}$.

35. $11\frac{1}{4}$.
36. $8\frac{7}{8}$.
37. $25\frac{7}{8}$.
38. $17\frac{9}{10}$.
39. $25\frac{7}{10}$.
40. $10\frac{7}{8}$.
41. $23\frac{9}{10}$.
42. $19\frac{3}{8}$.
43. $69\frac{1}{16}$.
44. $3\frac{11}{14}$.
45. $8\frac{7}{24}$.
46. $7\frac{1}{10}$.
47. $5\frac{17}{20}$.
48. $106\frac{5}{16}$.
49. $12\frac{11}{12}$.
50. $7\frac{11}{12}$.
51. $2\frac{3}{4}$.
52. $45\frac{5}{8}$.
53. $25\frac{1}{10}$.
54. $38\frac{9}{16}$.
55. $21\frac{9}{10}$.
56. $3\frac{11}{12}$.
57. $18\frac{17}{18}$.
58. $1\frac{13}{14}$.
59. $7\frac{9}{10}$.
60. $15\frac{11}{12}$.

1. $7\frac{1}{12}$ yards.
2. $57\frac{1}{2}$ gallons.
3. 26 yards.
4. 31 cents.
5. $\$2.60$.

Page 222.
6. $\$1.40$.
7. $\$2.00$.
8. $13\frac{7}{8}$ pounds.
9. $\$9.64$.
10. $\$20$.
11. 360.
12. $\$36$.
13. $\$3.40$.
14. $\$48$.
15. $\$9.30$.

Page 223.
16. 16 marbles.
17. 20 stamps.
18. $64\frac{1}{2}$ yards.
19. 35 cents.
20. 2 days.

1. 258.
2. 2449.
3. 228.
4. 568.
5. 124.
6. 9920.
7. 339.
8. $199\frac{1}{2}$.
9. $293\frac{1}{2}$.
10. $1231\frac{1}{2}$.
11. $31\frac{1}{2}$.
12. $14\frac{3}{8}$.
13. $14\frac{1}{4}$.

ANSWERS.

14. $14\frac{1}{8}$.
15. $16\frac{1}{8}$.
16. $21\frac{2}{13}$.
17. $50\frac{1}{10}$.
18. $45\frac{1}{2}$.
19. $75\frac{2}{25}$.
20. $144\frac{1}{19}$.
21. $246\frac{1}{11}$.
22. $302\frac{1}{15}$.

Page 225.
1. 225,506,736.
2. 804,580,398.
3. 561,276,891.

Page 226.
1. $609,340.37.
2. $680,494.41.
3. $840,200.33.

Page 227.
4. 350,879,581.
5. 627,020,401.
6. 589,140,749.
7. 668,386,689.
8. 777,993,982.
9. 713,200,695.
10. 578,616,033.

Page 228.
11. 65,461,219.
12. 615,808,906.
13. 99,090,910.
14. 200,290,240.
15. 249,054.
16. 26,081.
17. 102,900,999.

18. 31,660,868.
19. 82,816,981.
20. 6,543,211.
21. 3,264,973.
22. 53,386,521.

───

1. $2,706,230.50.
3. 120,263,455 miles.

Page 229.
4. $608.
5. $18.

Page 230.
3. 12,642,968.
4. 8625.
5. 980,304.
6. $439.11.
7. $314.87.
8. 7225.
10. $55,350.
11. 1,207,053.
13. 1376 yards.
14. 998,392.

Page 231.
25. $12\frac{3}{8}$.
26. $19\frac{1}{4}$.
27. $42\frac{1}{4}$.
28. $79\frac{1}{4}$.
29. $80\frac{3}{8}$.
30. $128\frac{4}{5}$.
31. $101\frac{1}{8}$.
32. $169\frac{1}{10}$.
33. $215\frac{4}{11}$.
34. $177\frac{1}{12}$.
35. $111\frac{5}{8}$.

36. $96\frac{1}{4}$.
37. $163\frac{1}{8}$.
38. $266\frac{1}{8}$.
39. $245\frac{5}{8}$.
40. $348\frac{4}{5}$.
41. $244\frac{1}{8}$.
42. $378\frac{1}{10}$.
43. $468\frac{4}{11}$.
44. 257.
45. $309\frac{1}{13}$.
46. $155,090\frac{1}{4}$.
47. $6,108,538\frac{4}{5}$.
48. $3,761,048\frac{4}{7}$.
49. $25,011\frac{2}{3}$.
50. $28,508\frac{1}{4}$.
51. $69,763\frac{2}{3}$.
52. $598,686\frac{1}{5}$.
53. $3,600,925\frac{5}{8}$.
54. $21,436,213\frac{1}{8}$.

Page 232.
1. $18,016.14.
2. 1,058,213.
3. $3405.78.
4. 8502.
5. 21,263,502.
6. $747\frac{3}{5}$.
7. $22,432\frac{129}{512}$.

Page 233.
3. 19,656.
4. 381.
5. $62.50.

───

1. 674,022,122 pieces.
2. $2,466,338.49.
3. 38,788.
4. $9332.86.

5. 75 pounds.
6. 1000.

Page 234.
8. $672.
$1.20.

Page 235.
1. 6.
2. 12.
3. 12.
4. 24.
5. 4.
11. 5.
12. 15.
13. 13.
14. 31.
15. 5.
16. 11.
17. 17.
18. 25.
19. 5.
20. 8.
31. 16.
32. 8.
33. $20\frac{1}{4}$.
34. $30\frac{1}{4}$.
35. 10.

Page 236.
41. 24.
42. 40.
43. 17.
44. 61.
45. 9.
46. 90.
47. 6.

ANSWERS.

1. $1283.35.
2. $845.95.
3. $2121.75.
4. $857.62.
5. $1247.80.
6. $769.89.
7. $530.25.
8. $853.58.
9. $1250.93.
10. $712.50.

Page 237.
1. $1115.02.
2. $505.44.
3. $1592.64.
4. $9263.05.
5. $1526.25.
6. $967.20.
7. $7133.80.
8. $1072.56.
9. $23.76.
10. $58.24.

Page 238.
11. $92.88.
12. $989.90.
13. $102.52.
14. $133.38.
15. $140.60.
16. $34.98.
17. $53.07.
18. $103.60.
19. $1591.62.
20. $4879.77.

1. 4,680,785$\frac{1}{2}$.
2. 2,478,208.
3. $258,715,000.
4. 4919.

5. 80,191.
6. 74 acres.
7. 46,200$\frac{1}{17}$.
8. 18 patterns.

Page 239.
9. 2582$\frac{231}{334}$.
10. $89.60.
13. 5 quarts.
14. 71 quarts.
15. 4 pecks.
16. 9.
17. $2.10.
18. 49,275.

Page 246.
1. 40 days.
2. 1050 yards.
3. 25 sheep.
4. 149 pounds.
5. 300 bushels.
6. 1402 pounds.
7. 36 cents.
8. 66$\frac{2}{3}$ cents.

Page 247.
9. 522 rabbits.
10. $2837.50.
11. $11,159.
12. 24 horses.
13. 1000 bars.

Page 249.
5. 915$\frac{431}{444}$.
6. 16 days.
7. 19,208 lb.
8. 57 inches.

1. $86,362$\frac{380}{1817}$.
2. 1288 pieces.
3. $566.

4. $378.75.
5. 41 days.
6. 5578$\frac{1}{2}$ bu.
7. 148$\frac{1}{18}$ gal.

Page 250.
8. 1,268,459,565 pounds.
9. 1,125,025,619 yards.
10. 30 hours.
12. 2,111,310,206.
13. 10,209,990 pieces.
14. 3684 quarts.
15. $1645.56.

Page 251.
1. 180 hours.
2. 180 hours.
3. 7200 seconds.
4. $12.75.
5. $2.48.
6. 40 pints.
7. 160 hf. pt.
8. 20 packages.
9. 25 cents.

Page 252.
10. 6 gallons.
11. 18,000 sec.
12. 10,080 min.
13. 800 minutes.
14. 512 quarts.
15. 757 ounces.
16. 14 lb. 13 oz.
17. 21 hr. 54 min.
18. 1008 hours.
19. 744 hours.
20. 98 inches.
21. 128,000 oz.

22. 9280 ounces.
23. 750 pounds.
24. 15 cwt.
25. $1.
26. $\frac{3}{4}$ ton.
27. 4 da. 16 hr.
28. 50 yards.
29. $8.25.
30. 32 cents.
31. $\frac{3}{4}$ cwt.
32. 348 pints.

Page 253.
4. 39,312.
   24,568.
5. 152$\frac{4}{18}$.
   69$\frac{111}{222}$.

Page 258.
1. $134,083.44.
2. $108,350.78.
3. $27,437.70.
4. $56,284.66.
5. $11,672.66.
6. $96,229.43.

Page 259.
7. 1,000,342.
8. 854,822.
9. 3,649,094.
10. 11,810,804.
11. 77,472,965.
12. 33,845,968.
13. 27,749,898.
14. 28,338,290$\frac{2}{3}$.
15. 32,136,750.
16. 6,248,365$\frac{1}{2}$.
17. 29,654,230.
18. 63,257,616$\frac{2}{3}$.
19. 857,375.
20. 274,170.

ANSWERS.

21. 72,243.
22. 109,243,616.
23. 115,242⅘.
24. 16,610,750.

Page 260.
25. 93₃⁶⁹⁄₇₀.
26. 2175.
27. 1025¹¹⁄₂₇.
28. 296₇₂₁⁰⁰⁰.
29. 530¹⁴⁹⁄₁₅₄.
30. 6943₃⁶⁰¹⁰²⁴.
31. 5565₁⁹²⁷³⁸⁴⁰⁵.
32. 15,168²⁸⁵⁷⁹³⁸.
33. 2708.
34. 920²⁸⁴⁹⁄₄₀₅₈.
35. 870³⁸⁹⁹⁄₈₀₃₃.
36. 216¹⁹⁄₇₀⁸⁵⁵.

1. 39.
2. 11.
3. 12.
4. 88.
5. 181.
6. 77.
7. 690.
8. 17.
9. 11.
10. 123.

Page 261.
11. 13.
12. 20.

Page 262.
1. $3.
2. $3.
3. $550.
4. $1120.
5. $105.
6. 5 cents.

7. $1536.65.
8. 49 T. 820 lb.
9. 22.
10. $80.
11. $1.25.
12. 25 cents.
13. 60 yards.

Page 263.
14. $22.
15. 10 papers.

———

1. $12.04.
2. $19.63.
3. $107.52.
4. $8.29.

Page 264.
5. $122.75.

Page 266.
1. 34,876¼ average applications. $245,468.65⅕ average surplus.
5. $4.45⅜.
6. 50₁₉⁹⁵⁰₁ cts.

Page 267.
5. 122.995.
6. 293.056.
7. 59.556.
8. 404.529.
9. 390.732.
10. 300.417.
11. 480.507.
12. 939.186.
13. 1180.106.
14. 104.231.

15. 23.495.
16. 18.168.
17. 2359.925.
18. 748.311.
19. 1062.556.
20. 799.511.

Page 269.
41. 1.08.
42. 400.4.
43. 780.8.
44. 780.8.
45. 2.68.
46. 1.536.
47. 1.536.
48. 55.272.
49. .485.
50. 4344.
51. 330.
52. 960.
53. 18.
54. 3801.
55. 59.13.
56. 725.56.
57. 376.68.
58. 334.508.
59. 62.7.
60. 1.8.

Page 271.
31. $241.25.
32. 1968.5 yd.
33. 35.4 pounds.
34. 1212.5 lb.
35. 2.64 tons.
36. $3364.02.
37. 19 pints.
38. .125 peck.
39. $58.50.
40. $1751.56¼.

Page 273.
1. 6.375 yards.
2. $95.386.
3. 2060³⁄₇ acres.
4. $14,000.
5. 67 yards.

Page 274.
6. 60 gills.
7. 12 bushels.
8. 26₁⁷₈ gallons.
9. $544.
10. $1960.
13. $4.50 gain.
14. $1.76.
15. 704 pints.
16. $679.
17. $15.66 gain.
18. $2.16.
20. 8 marbles.
21. 90 cents.
22. 88 quarts.

Page 275.
23. 90 cents.
24. 558 pupils.
25. 110 feet.
26. $1806.
27. 20 cents.
28. $2.48.
29. 491 gills.

———

4. 21¹⁸⁄₂₅.
5. 7³³⁄₅₆.
6. 95 cents.

Page 276.
7. $1.25.
8. 22¼ yards.
9. $80.50.
10. 1050 hours.

ANSWERS. 5

11. 33¾ hours.
13. 2⅛ miles.
14. $148.03.
15. 7½ weeks.
16. $1.87½.
17. $2.02.
18. 254 hf. pt.
19. 9717⅞⅞.

Page 277.
1. 182 sq. in.
2. 153 sq. in.
3. 126 sq. in.
4. 345 sq. in.
13. 180 sq. in.
14. 144 sq. in.
15. 192 sq. in.
16. 360 sq. in.
17. 450 sq. in.
18. 1419 sq. in.
19. 1130 sq. in.
20. 2205 sq. in.

Page 278.
1. 168 sq. ft.
2. 255 sq. ft.
3. 209 sq. ft.
4. 345 sq. ft.
5. 288 sq. ft.
6. 348 sq. ft.
7. 186 sq. ft.
8. 186 sq. ft.
9. 300 sq. ft.
10. 300 sq. ft.
11. 423 sq. ft.
12. 444 sq. ft.
13. 308 sq. ft.
14. 426 sq. ft.
15. 386 sq. ft.
16. 843 sq. ft.

17. 321¼ sq. ft.
18. 320 sq. ft.
19. 400 sq. ft.
20. 150 sq. ft.

Page 279.
1. 10,937.5 sq. ft.
2. 2¾ sq. yd.
3. 4½ sq. yd.
4. 117 sq. m.
5. $15.
6. $48.75.
7. $2.40.
8. 672 sq. rods.
9. 30 sq. yd.
10. 4 sq. yd.

Page 280.
1. 22$\frac{7}{4}$.
2. 12$\frac{44}{168}$.
3. 18$\frac{18}{28}$.
4. 48$\frac{887}{884}$.
5. 47$\frac{1}{10}$.
6. 90$\frac{11}{15}$.
7. 3$\frac{1}{20}$.
8. 48$\frac{18}{19}$.
9. 103$\frac{2}{3}$.
10. 99$\frac{11}{15}$.

Page 281.
11. 8$\frac{11}{12}$.
12. 6$\frac{11}{14}$.
13. 38$\frac{3}{5}$.
14. 17$\frac{12}{13}$.
15. 19$\frac{2}{3}$.
16. 16$\frac{11}{18}$.
17. 24$\frac{17}{10}$.
18. 22$\frac{7}{8}$.
19. 6262$\frac{17}{11}$.
20. 199$\frac{11}{17}$.

21. 74.
22. 3$\frac{5}{17}$.
23. 16$\frac{17}{14}$.
24. 318.
25. 4$\frac{1}{10}$.
26. 22$\frac{2}{3}$.
27. 46$\frac{3}{7}$.
28. 2437.
29. 507$\frac{7}{8}$.
30. $\frac{2}{45}$.

Page 282.
1. 2, 43.
2. 3, 29.
3. 2, 2, 2, 11.
4. 2, 3, 3, 5.
5. 7, 13.
6. 2, 2, 23.
7. 3, 31.
8. 2, 47.
9. 5, 19.
10. 2, 2, 2, 2, 2, 3.
11. 2, 2, 5, 5.
12. 2, 2, 2, 3, 5.
13. 2, 3, 5, 7.
14. 2, 2, 2, 2, 3, 5.
15. 2, 2, 2, 3, 3, 5.
16. 2, 2, 2, 2, 2, 2, 3, 3.
17. 2, 2, 2, 3, 5, 7.
18. 2, 2, 2, 2, 2, 2, 2, 3, 3.
19. 2, 2, 2, 2, 2, 3, 3, 3.
20. 2, 2, 2, 2, 2, 3, 3, 7.

Page 283.
1. $\frac{2}{3}$.
2. $\frac{11}{14}$.
3. $\frac{7}{8}$.
4. $\frac{3}{4}$.
5. $\frac{11}{14}$.
6. $\frac{1}{4}$.
7. $\frac{27}{37}$.
8. $\frac{17}{27}$.
9. $\frac{3}{5}$.
10. $\frac{3}{4}$.
11. $\frac{23}{25}$.
12. $\frac{3}{4}$.

Page 284.
3. $\frac{3}{5}$.
4. $\frac{11}{15}$.
5. $\frac{1}{4}$.
6. $\frac{7}{13}$.
7. $\frac{3}{5}$.
8. $\frac{7}{8}$.
9. $\frac{4}{17}$.
10. $\frac{3}{5}$.
11. $\frac{7}{13}$.
12. $\frac{3}{4}$.
13. $\frac{7}{11}$.
14. $\frac{4}{5}$.
15. $\frac{9}{10}$.

Page 286.
1. 120.
2. 900.
3. 840.
4. 240.
5. 600.
6. 48.
7. 360.
8. 231.
9. 720.
10. 420.

ANSWERS.

1. $16\frac{4.3}{5.5}$.
2. $84\frac{1.1}{6.0}$.
3. $66\frac{8.3}{1.0.0}$.
4. $31\frac{5}{8}$.
5. $61\frac{7}{1.0}$.
6. $2\frac{4.7}{3.0.0}$.
7. $82\frac{1.9}{6.0}$.
8. $101\frac{6.2.7}{1.0.0.0}$.
9. $41\frac{9.1}{1.2.0}$.
10. $23\frac{7.8}{1.2.5}$.

Page 287.
11. $8\frac{7}{4.8}$.
12. $9\frac{1}{2.4}$.
13. $85\frac{9.7}{1.4.4}$.
14. $74\frac{2.7}{3.0}$.
15. $79\frac{2.9}{3.0}$.
16. $223\frac{9.4}{1.0.5}$.
17. $471\frac{6.5}{1.0.8}$.
18. $10\frac{9.6.7}{1.0.0.0}$.
19. $43\frac{1.3}{3.8}$.
20. $9\frac{4.9}{1.0.0.0}$.
21. $\frac{2}{7.5}$.
22. $\frac{1}{2.5}$.
23. $17\frac{7}{2.0}$.
24. $6\frac{1.1}{4.0}$.
25. $13\frac{1}{3}$.
26. $674\frac{1}{2}$.
27. $1\frac{1}{1.8}$.
28. 3.
29. $1036\frac{5}{7}$.
30. $8\frac{8.6}{1.0.5}$.

Page 289.
1. $19\frac{1}{4}$ cents.
2. 75 cents.
3. $416\frac{1}{4}$ miles.
4. 2.
5. $5.
6. $107.48.
7. $\frac{3}{4}$.

8. 532 bags.

Page 290.
9. 52.272 acres.
10. 26,100 sec.
11. $6.35.
12. $7.10.
13. $20\frac{5}{8}$ yards.
14. 50 cts. 15. $\frac{1}{5}$.
16. 68 marbles.
17. 80 cents.
18. $119.46.
19. $58\frac{1}{4}$ miles.
20. $6.90.

Page 293.
1. $581,812,541,-985.20.
2. 3,417,600 acres.

Page 294.
3. $2480.
6. 156 pounds.
7. $422.
8. 14,688 ounces.
10. 3 cents.

1. $4,532,088.68.

Page 295.
2. 2551 gal. 1 qt. 1 pt.
3. $25.74 lost.
6. $524,470,971.-05.
7. 3.33\frac{1}{3}$.
8. $2600.
9. $1017.25.

Page 296.
1. 135.

2. $105\frac{4.4}{5.1}$.
3. $117\frac{3.9}{5.1}$.
4. $121\frac{5.2}{8.4}$.
5. $219\frac{5.4}{5.7}$.
6. $344\frac{7.5}{1.2.3}$.
7. $213\frac{4.9.5}{3.4.5}$.
8. $168\frac{4.4.9}{5.4.9}$.
9. $420\frac{5.9.8}{1.0.0.8}$.
10. $342\frac{6.6.5}{2.5.3.4}$.

Page 297.
3. 36 days.
4. 68 bags.
5. $2.99.
6. $78.
7. $2844.
8. $1.50.
9. $12.
10. $147.60.
11. $704.
12. 36 days.

Page 298.
13. $2.31.
14. 60 days.
15. $87\frac{1}{2}$ cents.

Page 299.
1. 64.
2. 96.
3. $\frac{1.2}{3.5}$.
4. $\frac{3.1}{4.1}$.
5. $\frac{1}{2}$.
6. 246.
7. 436.
8. $2221\frac{1}{5}$.
9. $3115\frac{7}{1.5}$.
10. 15.
11. 9.
12. $1\frac{1}{2}$.
13. 115.

14. $3\frac{1}{2}$.
15. $37\frac{3}{4}$.
16. 3.
17. 42.
18. $3\frac{2.3}{2.5}$.
19. $\frac{1}{4}$.
20. $\frac{1}{2}$.
21. $\frac{9}{2.5}$.
22. $\frac{8}{2.7}$.
23. $57\frac{7}{8}$.
24. $41\frac{1}{7}$.
25. $47\frac{1.3}{1.8}$.
26. $22\frac{1}{1.0}$.
27. $10\frac{1}{5}$.
28. $9\frac{1}{1.1}$.
29. $11\frac{1}{3}$.
30. $15\frac{5}{8}$.
31. $3\frac{1}{4}$.
32. $\frac{2.7}{4.7}$.
33. $\frac{1.4}{1.8}$.
34. $5\frac{1}{1.5}$.
35. $5\frac{1.7}{1.4.4}$.
36. $178\frac{7}{8}$.
37. $19\frac{1.1}{1.2}$.
38. $7\frac{5.8}{6.9}$.
39. $2\frac{3}{5}$.
40. $1\frac{3}{1.0}$.

Page 301.
1. 10.
2. $\frac{1}{4}$.
3. $23\frac{1}{3}$.
4. $3\frac{3}{4}$.
5. $\frac{7}{7.0}$.
6. $\frac{4}{1.5}$.
7. $\frac{7}{2.0}$.
8. $2\frac{4}{9}$.
9. $7\frac{1.1}{1.3}$.
10. $\frac{4.9}{1.8.7}$.
11. $1\frac{5}{1.4}$.
12. $\frac{3}{4}$.

13. $1\frac{1}{3}$.
14. $2\frac{3}{5}$.
15. $\frac{3}{4}$.
16. $\frac{4}{11}$.
17. 6.
18. $\frac{1}{6}$.
19. $\frac{1}{17}$.
20. $\frac{11}{197}$.
21. $\frac{4}{9}$.
22. $\frac{8}{15}$.
23. $3\frac{11}{16}$.
24. $2\frac{1}{16}$.
25. $2\frac{139}{175}$.
26. $2\frac{98}{171}$.
27. $1\frac{41}{70}$.
28. $1\frac{97}{735}$.
29. $1\frac{7}{32}$.
30. $3\frac{1}{2}$.
31. $6\frac{1}{4}$.
32. $4\frac{1}{6}$.
33. $17\frac{1}{7}$.
34. $17\frac{1}{7}$.
35. $30\frac{19}{21}$.
36. $30\frac{19}{21}$.
37. $5\frac{3}{5}$.
38. $33\frac{9}{10}$.
39. $37\frac{1}{2}$.
40. $6\frac{2}{3}$.
41. $101\frac{13}{115}$.
42. $2\frac{359}{1008}$.

Page 305.

7. $1\frac{5}{6}$.
8. \$ $1\frac{23}{45}$.
9. $10\frac{1}{2}$.
10. 4.
11. $59\frac{2}{3}$.
12. \$ $1\frac{43}{60}$.
13. $1\frac{19}{35}$.
14. $1\frac{1}{24}$.
15. $4\frac{64}{175}$.

Page 307.

1. \$ $2.43\frac{1}{2}$.
2. 1 ft. 3 in.
3. $10\frac{1}{8}$ pounds.
4. $\frac{7}{72}$.
5. 60 cents.
6. 96 hours.
7. \$ 1.14.
8. 30 days.
9. \$ 76.25.

Page 308.

10. 32 days.
11. 6 yards.
12. 1,260,000 cu. ft.
13. \$ 7.50.
14. 90 cents.
15. 964 pounds.
16. 60 cents.
17. $31\frac{1}{4}$ cents.
18. $\frac{16}{45}$.
19. \$ 161.
20. Twice as old.
21. $37\frac{3}{5}$.
22. \$ 1430.
23. 2 days.
24. $\frac{35}{103}$.
25. \$ 160.

Page 311.

1. \$ 161.85.
2. \$ 27.36.
3. \$ 35.96.
4. \$ 54.95.

Page 312.

5. \$ 78.27.

---

3. \$ 882,258.55.
4. \$ 4,211,587.67.

Page 313.

4. 53 hours 20 minutes.
5. \$ 4044.63.
\$ 2.36.
\$ 2.26.
\$ 2.15.
6. \$ 5,400,000.

---

3. 930,435,246 pieces.
4. 10,209,990 pieces.
7. \$ 320.
8. \$ $1.33\frac{83}{117}$.

Page 314.

1. 29 pupils.
2. $\frac{4}{5}$.
4. \$ $255.93\frac{3}{4}$.
5. 8.
6. $2\frac{11}{60}$.
7. $\frac{2}{3}$.
8. $\frac{7}{9}$.
9. $16\frac{8}{15}$.
10. \$ 45.

Page 315.

Carriages, 2,698,526.
Equestrians, 132,137.
Pedestrians, 13,730,597.
Total, 16,567,956.

Page 318.

1. 80.
2. 80.
3. 28.8.
4. 27.
5. 72.96.
6. 10.8.
7. 81.666.
8. 47.25.
9. 24.
10. 124.962.

Page 319.

11. 112.
12. 335.
13. 445.9.
14. 120.
15. 26.748.
16. 372.6.
17. 473.184.
18. 222.
19. 10.92.
20. 15,701.57.
21. 32.
22. 3.2.
23. 700.
24. 40.
25. 40.
26. 2.
27. 3.2.
28. 30.6.
29. 200.
30. 144.
31. 32,000.
32. 121.
33. 13,500.
34. 12.5.
35. 42.1.
36. 15,100.
37. 1510.
38. 151.

Page 320.

39. 21.

ANSWERS.

40. 21.
41. 21.
42. 21.
43. 240.
44. 300.
45. 3.5.
46. 12,360.
47. 122.5.
48. .016.
49. 400.
50. .007.
51. 47.
52. .064.
53. 13.5.
54. 43647.89+.
55. 2.384+.
56. .264+.
57. 20.001+.
58. 4.405+.
59. 24.8.
60. 2.634+.

Page 321.
4. .812.
5. .105.
6. 1.06.
7. .143.
8. .025.
9. .044.
10. .05625.
11. .105.
12. .288.
13. 36.4.
14. .088.
15. .605.

Page 323.
1. $562.68.
2. 105.45 sq. rods.
3. 14 rods.

4. 420.168+ marks.
5. 103.771.
6. 2699.73.
7. 77.7.
8. .3.
9. 1864 lb.
10. 62.832 in.
———
3. 61.
4. 540.
6. $7\frac{3}{4}$.
7. $4\frac{13}{16}$ pounds.

Page 324.
8. $\$\frac{11}{12}$.
9. 12 days.
10. 14 yr. 1 mo.
11. 6 miles.
12. $1.92.
13. $87\frac{1}{2}$.
    $3\frac{9}{10}$.
14. $5.86.

Page 325.
2. 5,026,101.
3. 1,359,908.
4. $9319\frac{113}{367}$.
5. $2\frac{2}{3}$ yards.
6. $12.40.
7. 1 mile.
8. $2.58.

Page 327.
1. 17 days.
2. $\frac{1}{4}$ week.
3. $216.
4. 5400 min.
5. 5 da. 15 hr.
6. 18 hours.
7. 9 hr. 36 min.

8. $28.
9. 82 cents.
10. 4 lb. 10 oz.
11. 6 bu. 1 pk. 5 qt.
12. 4 gallons. 1$\frac{1}{2}$ pints.
13. $3.05.
14. 9 bu. 1 qt.

Page 328.
15. 63,360 in.
16. 924 feet.
17. 14,080 rails.
18. 3840 steps.
19. 55 minutes.
20. 15 hr. 17 min. 14 hr. 18 min.
21. 276 ounces.
22. 169,560 lb.
23. 151 quarts.
24. 360 pints.
25. 127 pints.
26. 111 pecks.
27. 391 quarts.
28. 1344 pints.
29. 47,520 yd.
30. 91 yards.
31. 10 da. 10 hr.
32. 12 T. 1124 lb.
33. 100 rods.

Page 329.
34. 109 gal. 2 qt.
35. 6 yd. 1 ft.
36. 5 mi. 50 rd.
37. 5 wk. 1 da.
38. 9 bu. 1 pk.

39. 19 lb. 11 oz.
40. 61 yr. 11 mo.
41. 25 ft. 2 in.
42. 105 min. 57 sec.
43. 50 years.
44. 35 wk. 2 da.
45. 13 miles.
46. 60 yards.
47. 50 gal. 2 qt.
48. 13 pk. 2 qt.
49. 74 bu. 1 pk.
50. 50 quarts.
51. 50 quarts.
52. 74 bu. 1 pk.
53. 48 pk. 3 qt.
54. 146 gallons.
55. 200 yards.
56. 63 mi. 175 rd.
57. 77 wk. 1 da.
58. 72 yr. 6 mo.
59. 81 min. 10 sec.
60. 113 feet.
61. 6 ft. 5 in.
62. 22 min. 43 sec.
63. 6 yr. 2 mo.
64. 31 wk. 5 da.
65. 6 mi. 177 rd.

Page 330.
66. 14 yd. 2 ft.
67. 145 gal. 3 qt.
68. 24 pk. 2 qt.
69. 84 bu. 3 pk.
70. 68 quarts.
71. 43 qt. 1 pt.

ANSWERS.

72. 10 min. 7 sec.
73. 17 yr. 5 mo.
74. 24 wk. 3 da.
75. 7 mi. 65 rd.
76. 25 yd. 1 ft.
77. 64 gal. 2 qt.
78. 24 bu. 3 pk.
79. 37 qt. 1 pt.
80. 9 bu. 1 pk. 4 qt.
81. 2.
82. 5.
83. 8.
84. 9.
85. 10.

1. 378 sq. yd.
2. 378 sq. yd.
3. 6 sq. yd.
4. 893 sq. yd.
5. 5963 sq. yd.
6. 396 sq. yd.
7. 288 sq. yd.
8. 36 sq. yd.
9. 240 sq. yd.
10. 36 sq. yd.

Page 331.
13. 420 sq. in.
14. 32 sq. yd.
15. 256 sq. yd.
16. 1500 sq. ft.
17. 270 sq. ft.

Page 332.
18. 96 sq. in.
19. 72 cents.
20. $40.

Page 334.
1. 10,061,280 minutes.
2. $6336.12.
3. 7609 cords.
4. $6934.89+.
5. $1833.72.
6. $70.20.
7. $608.
8. $101,790.
9. 3192; 45.
10. $1200.
11. 1.955 yards.
12. 7960.
13. 2.35\frac{2}{3}$.

Page 335.
14. $2.56.
15. 18 pounds.
16. 327 feet.
17. $17.88.
18. 62 days.
19.
20. $2.81.
21. $2.
22. 48 geographies.
23. 96.

Page 336.
24. $340.30.
25. 21 bushels.
26. 11,608 bot.
27. 192 pounds.
28. $72.06.
29. 6.
30. 60.

Page 337.
1. .00125.
2. .025.
3. .08.
4. .78125.
5. .265625.
6. .1875.
7. .006.
8. .03125.
9. .00525.
10. .0035.
11. .24.
12. .056.
13. 2.875.
14. 2.9375.
15. .044.
16. .0016.
17. 5.859375.
18. .1015625.
19. .00390625.
20. .013671875.
21. .0009765625.

Page 338.
22. $\frac{1}{100}$.
23. $\frac{9}{25}$.
24. $\frac{11}{100}$.
25. $1\frac{1}{4}$.
26. $\frac{3}{37}$.
27. $\frac{244}{333}$.
28. $\frac{4}{5}$.
29. $\frac{4}{5}$.
30. $\frac{18}{125}$.
31. $\frac{3}{5000}$.
32. $\frac{27}{100}$.
33. $\frac{27}{1000}$.
34. $\frac{7}{20000}$.
35. $\frac{24}{25}$.
36. $\frac{7}{10000}$.
37. $1\frac{23}{125}$.
38. $\frac{3}{37}$.
39. $\frac{3}{10}$.
40. $\frac{1}{100}$.
41. $\frac{12}{125}$.
42. 304.134.
43. 1167.3946.
44. 883.2429.
45. 2759.5755.
46. 2283.9171.
47. 314.7032.
48. 1013.0418.
49. 639.7105.
50. 49.21619.
51. 563.7625.

Page 339.
52. 188.26.
53. 288.3623.
54. 999.999.
55. 13.615.
56. 184.7569.
57. 15.0885.
58. 1999.96875.
59. 113.1991.
60. 17.84375.
61. 1503.5975.
62. 79.2.
63. .045264.
64. 1850.3125.
65. 4.566.
66. .13875.
67. .009438.
68. 6.3784.
69. 16.93542.
70. .45953125.
71. 45.78644.
72. 25.327+.
73. 81519.856+.
74. .222+.
75. .321+.
76. 88.4507+.
77. 23.328+.
78. 2626.595+.
79. .2025+.
80. .0655+.
81. 16.841+.

ANSWERS.

82. 544.382+.
83. 9.245+.
84. 5.343+.
85. .0438+.
86. 60.331+.
87. 304.977+.
88. 15.472+.
89. 17.426+.
90. 74.3802+.
91. 88.537+.

Page 340.
1. .034375.
2. 1200.
3. $407,294\tfrac{1030}{1734}$.
4. $70,234,730,841$.

Page 341.
5. $444.75.
6. 6.
7. .2955.

---

1. $152.50.
2. $7.22.
3. $136.08.
4. $836.02½.
5. $12.75.
6. $2392.39.
7. $111.45.
8. $26.23¼.
9. $157.50.
10. $31.50.
11. $579.72.
12. $47.41.
13. $546.48.
14. $1129.11.
15. $644.62.

Page 342.
16. $433.07.
17. $1787.95.

18. $125.66.
19. $1580.25.
20. $35.40.
21. 9.96.
22. .0002075.
23. 2.292.
24. 26.
25. 9.2.
26. 900.
27. 5.67.
28. .01008.
29. .33375.
30. .04375.
31. $\tfrac{1}{4}$.
32. $\tfrac{1}{4}$.
33. $\tfrac{1}{8}$.
34. $\tfrac{1}{7}$.
35. $\tfrac{1}{160}$.
36. $\tfrac{1}{90}$.
37. $\tfrac{1}{15}$.
38. $\tfrac{5}{8}$.
39. $\tfrac{1}{110}$.
40. $\tfrac{1}{24}$.
41. $\tfrac{10}{13}$.
42. $\tfrac{3}{80}$.

Page 343.
4. 5229 miles.
5. ½ cent.
6. $1,303,095.17.
7. 38 clerks.

Page 344.
8. $1914.65.
9. 34,888 packages.
10. 21,781.53696.

Page 345.
1. 1512 sq. in.
2. 1278 sq. in.

3. 1850 sq. in.
4. 4788 sq. in.
5. 31,104 sq. in.
6. 14,256 sq. in.
7. 810 sq. in.
8. 5980 sq. in.
9. 6264 sq. in.
10. 8424 sq. in.
11. 432 sq. ft.
12. 12 sq. ft.
13. 432 sq. ft.
14. 12 sq. ft.
15. 14 sq. ft.
16. 12 sq. ft.
17. 14 sq. ft.
18. 437½ sq. ft.
19. 14 sq. ft.
20. 1755 sq. ft.
21. 450 sq. yd.
22. 20 sq. yd.
23. 108 sq. yd.
24. 15 sq. yd.
25. 18 sq. yd.
26. 24 sq. yd.
27. $5\tfrac{5}{27}$ sq. yd.
28. $3\tfrac{1}{24}$ sq. yd.
29. 7¼ sq. yd.
30. 90 sq. yd.

Page 348.
1. 270 sq. ft.
2. 5 lb. 14 oz.
3. $127.32.
4. 20 cents.
5. $\tfrac{23}{40}$.
6. $\tfrac{7}{45}$.
7. $\tfrac{12}{25}$.
8. $4581\tfrac{9}{11}$ sec.

Page 349.
9. 216.

10. 29.
11. $6000.
12. 6 hr. 49 min. 45 sec. A.M.
13. A, $3600; B, $2400.
14. $1.35.
15. 45 sq. yd.
16. Increased $\tfrac{1}{12}$.
17. $12.
18. 360 oranges.
19. $68.40.
20. $750.

Page 350.
1. $15,373.84.
2. $15,697.16.
3. $40,525.88.

349,129 pupils.

Page 351.
1. $4\tfrac{91}{34}$ bushels.
2. $13,691.16.
3. Tea, 3555⅝ lb.; coffee, 6000 lb.; sugar, $37,012\tfrac{4}{83}$ lb.; $4,080 remaining.
4. 68.81495 lb.
5. Lost $45.97½.
6. $94.51.
7. $70.20.
8. 24.75 tons.
9. 204.0278267.

Page 352.
1. 233,675.
2. 64,725.
3. 101,537½.
4. 216,100.

ANSWERS. 11

5. 1,015,375.
6. 2,336,750.
7. 23,367¼.
8. 701,025.
9. 432,200.
10. 243,300.
11. 428,400.
12. 80,250.
13. 185,100.
14. 129,000.
15. 230,400.
16. 21,100.
17. 525,500.
18. 145,312½.
19. 24,062½.
20. 1,828,500.

Page 354.
1. 107,136.
2. 604,665.
3. 96,145.
4. 494,312.
5. 473,484.
6. 191,597.
7. 410,896.
8. 1,297,479.
9. 347,332.
10. 1,301,234.
11. 113,542.
12. 73,350.
13. 132,790.
14. 110,808.
15. 101,085.
16. 852,120.
17. 73,072.
18. 325,815.
19. 167,892.
20. 304,856.
21. 169,344.
22. 212,175.
23. 710,046.

24. 301,392.
25. 474,300.
26. 385,600.
27. 1,497,300.
28. 2,300,400.
29. 324,000.
30. 2,984,800.

---

1. $54,659,-886.61.

Page 355.
2. $145,543,-810.71.
5. $69.75.
6. 12 pages.
7. $13,614.07.
8. ¼.

Page 358.
2. 42 gal. 2 qt.
3. 637 gal. 2 qt.
4. 25,000 times.
5. 9½ yards.
6. 1501 min.
7. 66 days.
8. 41 hr. 15 min.
9. 1440 steps.

Page 359.
10. 80 rods.
11. 1$\frac{7}{9}$ min.
12. 4 gal. 1 qt. 1 pt.
13. 8 hr. 43 min.
14. 2 bu. 3 pk. 5 qt.
15. 75 rods.

---

1. 109$\frac{2}{10}$.
2. 25$\frac{7}{11}$.
3. 146$\frac{17}{11}$.

4. 97$\frac{11}{18}$.
5. 206$\frac{143}{110}$.
6. 240$\frac{7}{16}$.
7. 152$\frac{9}{20}$.
8. 211$\frac{4}{15}$.
9. 829$\frac{11}{14}$.
10. 224$\frac{43}{50}$.

Page 360.
11. 18¼.
12. 75$\frac{11}{12}$.
13. 36$\frac{9}{50}$.
14. 30$\frac{17}{41}$.
15. 49$\frac{11}{12}$.
16. 42$\frac{11}{14}$.
17. 37$\frac{11}{13}$.
18. 68$\frac{121}{108}$.
19. 16$\frac{11}{13}$.
20. 228$\frac{7}{60}$.
21. 17$\frac{2}{3}$.
22. 3¼.
23. 210$\frac{1}{7}$.
24. 52$\frac{5}{8}$.
25. 15$\frac{5}{8}$.
26. 3$\frac{7}{20}$.
27. 248.
28. 84$\frac{5}{8}$.
29. 4$\frac{7}{9}$.
30. 83¼.
31. 1¼.
32. 3¾.
33. 1384$\frac{2}{3}$.
34. 19$\frac{11}{22}$.
35. 8$\frac{1}{17}$.
36. 2$\frac{41}{112}$.
37. 2$\frac{11}{14}$.
38. 5$\frac{11}{14}$.
39. $\frac{5}{11}$.
40. $\frac{5}{9}$.
41. 31$\frac{1}{15}$.
42. 5$\frac{1}{12}$.

43. 43$\frac{5}{8}$.
44. 5$\frac{7}{11}$.
45. 49$\frac{7}{37}$.
46. $\frac{29}{120}$.
47. $\frac{5}{6}$.
48. 3$\frac{2}{15}$.

Page 361.
2. $\frac{1}{320}$.
.046875.
3. .000000140028.
4. $872.87.
5. 6$\frac{1}{11}$.
6. .09375 bu.
7. 11¼ pounds.
8. $22,612.50.
9. $4.05.
10. $296.25.
11. $968.88.
12. $20.16.
13. $7335.
14. $\frac{231}{10000}$.
15. $\frac{1}{15}$.
16. 560.22 yards

Page 362.
17. $7.96.
18. .04.
19. $676.
20. $6.
21. Gained $8.
22. 21 clerks.
23. 1280 sheep.
24. 4 boxes.
25. $\frac{15}{18}$.
26. $45.
27. $1033.05.
29. 31¼ cents.

Page 363.
30. .0002009877.

ANSWERS.

31. 82.
32. 31 years.
33. 21 27/44.
34. $108.
35. 399 yr. 2 mo. 17 da.
36. 219 hats.
37. 7 years.
38. $999.
39. .00012.
40. $3.
41. $110.
42. 63 miles.

Page 364.
1. 777 ounces.
2. 190 yards.
3. 3520 yards.
4. 89 hours.
5. 1455 seconds.
6. 17,675 lb.
7. 180 quarts.
8. 600 pints.
9. 79 pecks.
10. 632 quarts.
11. 62 lb. 8 oz.
12. 62 lb. 8 oz.
13. 2 ft. 4 in.
14. 2 ft. 3 in.
15. 3 qt. 1 pt.
16. 2 qt. 1 pt.
17. 1 pk. 7 qt.

Page 365.
1. 47,789⅔.
2. (a) 14.75605; (b) 5999.25.
3. 598 bu. 3 pk.
4. 15⅔.
5. 4 weeks.
6. 34 cords.
7. $357.50.

Page 366.
1. 60 lb. 15 oz.
2. 11 yards.
3. 21 da. 13 hr.
4. 28 min. 14 sec.
5. 4 T. 1314 lb.
6. 123 gal. 1 qt. 1 pt.
7. 185 pk. 5 qt.
8. 46 bu. 1 pk.
9. 5 weeks.
10. 990 inches.

Page 367.
1. 44 lb. 9 oz.
2. 23 yd. 1 ft.
3. 14 hr. 14 min.
4. 26 min. 13 sec.
5. 4 yd. 2 ft. 10 in.
6. 28 gal. 2 qt.
7. 18 bu. 3 pk.
8. 4 pk. 2 qt.
9. 12 weeks.
10. 16 T. 904 lb.

1. 3 lb. 9 oz.
2. 5 yd. 2 ft.
3. 7 hr. 10 min.
4. 33 min. 45 sec.
5. 1 ft. 4 in.
6. 18 gal. 2 qt.
7. 21 bu. 3 pk.
8. 3 quarts.
9. 1 wk. 3 da.
10. 4 T. 912 lb.

Page 368.
1. 42 days.
2. 12 days.
3. 28 days.
4. 56 men.
5. 33 horses.
6. 18 lines.
7. 900 steps.
8. 3072 bricks.
9. 11 hours.
10. 77 cents.
11. 12 days.

Page 369.
1. 168,932.
5. 15 hr. 16 min. $21\frac{9}{11}$ sec.
6. $40.
7. 6.0625.

Page 370.
8. $130.

1. 37 lb. 5 oz.
2. 22 hr. 10 min.
3. 49 T. 835 lb.
4. 69 bu. 3 pk.
5. 14 wk. 2 d.
6. 21 yd. 2 ft.
7. 73 minutes.
8. 19 gal. 2 qt.
9. 22 feet.
10. 9 yards.
11. 4 lb. 9 oz.
12. 3 gal. 2 qt.

13. 6 hr. 27 min.
14. 5 bu. 1 pk.
15. 5 min. 13 sec.
16. 2 yd. 2 ft.
17. 1 ft. 11 in.
18. 8 T. 1234 lb.
19. 2 wk. 6 da.
20. 4 yd. 2 ft. 3 in.
21. 4.
22. 6.
23. 8.
24. 9.
25. 7.
26. 9.
27. 8.

Page 371.
28. 12.
29. 15.
30. 13.
31. 16.
32. 11.
33. 18.

1. 1125 sq. ft.
2. 432 sq. ft.
3. 48 sq. yd.
4. 12 sq. yd.
5. 13 sq. yd.
6. 44 sq. yd.
7. 7975 sq. ft.

Page 373.
1. $548.80.
2. $37.45.
3. $72.
4. $187.60.
5. $137.10.
6. 11 pupils.
7. $480.

ANSWERS.

Page 374.
8. $333.
9. 3 words.
10. 60 cents.

Page 375.
2. $1.33⅓.
3. 1550.
4. .000007.
5. 4.975.
6. 2633.0045.
7. $6.75.
8. 66 cents.
9. 4 quarts.
10. 160 acres.
11. 5½ hours.
12. ⅘.
13. $\frac{16}{100}$; $\frac{8}{50}$; $\frac{4}{25}$; 16%.

Page 376.
1. $27.56.
2. $31.40.
3. $18.99.
4. $45.52.

Page 378.
1. $11.46.
2. $29.10.
3. $18.77.
4. $11.37.
5. $21.87.
6. $7.47.
7. $3.99.
8. $22.26.
9. $5.08.
10. $7.46.
11. $87.99.
12. $6.30.
13. $13.42.
14. $64.97.

15. $2.55.
16. $56.25.
17. $49.02.
18. $3.99.
19. $95.02.
20. $96.58.
21. $23.20.
22. $189.
23. $568.
24. $225.
25. $589.60.
26. $51.30.
27. $62.40.
28. $320.
29. $133.25.
30. $52.92.
31. $13.50.
32. $2.75.
33. $55.

Page 379.
4. 37½ sq. yd.
5. 900 sq. ft.
6. 2 sq. ft.
7. 3¾ sq. ft.
8. 8⅔ sq. ft.
9. 13¼ sq. ft.
10. 8100 sq. ft.

Page 380.
11. 5062½ sq. ft.
12. 7½ sq. ft.
13. 750 sq. ft.
14. 61⅞ sq. ft.
15. 308½ sq. ft.

1. 38$\frac{9}{13}$ cents.
2. $10.04.
3. 18 yd. 2 ft.
4. 3520 rails.

5. 497$\frac{44}{105}$ min.
6. $247.50.
7. 135 pounds.
9. 23 tons.
10. 231 pints.

Page 381.
1. 121¼.
2. 210⅓.
3. 88⅓.
4. 331$\frac{1}{17}$.
5. 139$\frac{5}{13}$.
6. 118⅞.
7. 591$\frac{1}{17}$.
8. 382$\frac{4}{15}$.
9. 247$\frac{33}{34}$.
10. 263⅝.
11. 5$\frac{7}{15}$.
12. 13$\frac{3}{110}$.
13. 21⅙.
14. 81⅞.
15. 12$\frac{11}{15}$.
16. 11$\frac{12}{13}$.
17. 12⅞.
18. 72$\frac{5}{13}$.
19. 30$\frac{5}{13}$.
20. 40$\frac{31}{33}$.

Page 383.
1. 117$\frac{11}{13}$ yd.
2. $3000.
3. $27.56.
4. $244¼.
5. $5.83⅓.

Page 384.
6. $87.
7. $\frac{1}{100}$; ⅘.
8. $4.42.

9. 300.04;
 6.568¼.
10. 1$\frac{1}{12}$ oranges.
11. 502¼ days.
12. $5833.33⅓.

Page 385.
13. 1360 pounds.
14. 2178 feet.
15. $52.50.
16. $1.12.
17. $4.50.
18. $52¼.
19. $379.50.
20. 2880 tiles.
21. $15,000;
  $350.
22. ⅞.
23. $877.22.
24. 724 bushels.
25. $1828.50.

Page 386.
26. 31 pounds.
27. 5¼ miles.
28. $\frac{9}{35}$.
29. 3240 bushels.
30. $360.

Page 387.
1. 2⅔⅔.
2. 1$\frac{54}{100}$.
3. $\frac{7}{11}$.
4. 6¼.
5. 1⅛.
6. ⅞⅞.
7. 34$\frac{11}{11}$.
8. ⅞⅞.
9. 1.
10. 15.
11. 3.679+.

ANSWERS.

12. .005.
13. .004375.
14. 3.78.
15. 102.390561.
16. 19,700.

Page 388.
17. .125; 8.
18. 90.
19. 1.36.
20. .26285.

1. $\frac{3}{11}$.
2. $\frac{5}{16}$.
3. $\frac{141}{254}$.
4. $\frac{4}{5}$.
5. $\frac{14}{41}$.
6. $\frac{11}{15}$.
7. $\frac{3}{11}$.
8. $\frac{300}{401}$.
9. $\frac{7}{8}$.
10. $\frac{4}{5}$.
11. $\frac{7}{11}$.
12. $\frac{2172}{2265}$.

Page 389.
1. 131 pints.
2. 220 pints.
3. 128 pints.
4. 129 pints.
5. 279 pints.
6. 252 pints.
7. 77 pints.
8. 85 pints.
9. 217 pints.
10. 39½ pints.
11. 39 gal.
12. 19 gal. 3 qt.
13. 51 gal.
14. 162 gal. 3 qt.
15. 7 gal. 3 qt. 1 pt.
16. 34 gal. 2 qt. 1 pt.
17. 17 gal. 1 qt. 1 pt.
18. 42 gal. 3 qt.
19. 15 gal. 3 qt.
20. 88 gal. 3 qt. 1 pt.

Page 390.
21. 633 inches.
22. 7594 yards.
23. 2391 quarts.
24. 2507 ounces.
25. 2271 inches.
26. 611 quarts.
27. 192 feet.
28. 510 pints.
29. 102 quarts.
30. 34,369 lb.
31. 54,960 sec.
32. 827 hours.
33. 120 hours.
34. 165 yards.
35. 40 ounces.
36. 52 yd. 4 in.
37. 29 lb. 11 oz.
38. 22 bu. 3 pk. 1 qt.
39. 6 da. 35 min.
40. 2 T. 972 lb.
41. 3 mi. 12 rd.
42. 14 gal. 2 qt. 1 pt.
43. 2 hr. 38 min. 3 sec.
44. 27 bu. 1 pk. 5 qt.
45. 93 lb. 7 oz.
46. 10 yd. 1 ft. 1 in.
47. 54 gallons.
48. 2 mi. 236 rd.
49. 39 gal. 3 qt. 1 pt.
50. 2 miles.

Page 391.
1. .0015 T.
2. $\frac{5}{378}$ day.
3. 45 minutes.
4. 45 minutes.

Page 392.
5. .00625 day.
6. $76.87½.
7. 3 T. 1504 lb.
8. $46.87.
9. 14 T. 1244 lb.
10. $\frac{21}{36}$ yard.
11. .890625 bu.
12. 3 pk. 6 qt.
13. 75 cents.
14. $6.86.
15. 2 qt. 1½ pt.
16. $\frac{49}{150}$.
17. $\frac{174}{203}$.
18. 1 bu. 1 pk. 1 qt.
19. .885 day.
20. 5280 feet.

---

1. 46 lb. 7 oz.
2. 43 bu. 2 pk. 1 qt.

Page 393.
3. 34 yd. 2 ft. 6 in.
4. 40 da. 19 hr. 55 min.
5. 186 gal. 3 qt.
6. 22 hr. 30 min. 28 sec.
7. 18 T. 862 lb.
8. 9 wk. 1 da.
9. 53 mi. 294 rd.
10. 76 yr. 8 mo.
11. 866 T. 899 lb.
12. 140 lb. 3 oz.
13. 38 hr. 40 min. 2 sec.
14. 180 gal. 3 qt. 1 pt.
15. 137 yd. 7 in.
16. 194 mi. 183 rd.
17. 128 yr. 4 mo. 21 da.
18. 36 wk. 5 hr.
19. 22 hr. 24 min. 22 sec.
20. 17 bu. 4 qt.

Page 394.
21. 17 lb. 7 oz.
22. 3 bu. 2 pk. 5 qt.
23. 7 yd. 2 ft. 7 in.
24. 10 da. 9 hr. 20 min.
25. 3 gal. 2 qt. 1 pt.
26. 13 hr. 44 min. 30 sec.

ANSWERS.

27. 246 T. 1676 lb.
28. 11 wk. 16 hr.
29. 15 mi. 311 rd.
30. 11 yr. 3 mo.
31. 16 lb. 12 oz.
32. 8 bu. 3 pk. 6 qt. 1 pt.
33. 8 yd. 1 ft. 9 in.
34. 12 da. 23 hr. 45 min.
35. 56 gal. 1 pt.
36. 67 yr. 6 mo.
37. 42 mi. 245 rd.
38. 38 T. 546 lb.
39. 16 lb. 12 oz.

Page 395.
40. 38 wk. 3 da. 17 hr.
41. 10 gal. 1 qt. 1 pt.
42. 17 hr. 24 min. 35 sec.
43. 8 yd. 1 ft. 10 in.
44. 57 bu. 1 qt.
45. 38 da. 18 hr. 55 min.
46. 13 bu. 1 pk. 6 qt.
47. 16 gal. 2 qt. 1 pt.
48. 6 hr. 29 min. 40 sec.
49. 3 lb. 10 oz.
50. 25 bu. 1 pk. 7 qt.
51. 27 bu. 3 pk. 4 qt.
52. 76 gal. 3 qt. 1 pt.
53. 30 lb. 8 oz.

54. 16 hr. 16 min. 15 sec.
55. 138 bu. 2 pk. 2 qt.
56. 224 gal. 3 qt. 1 pt.
57. 202 pounds.
58. 5 hr. 1 min. 57 sec.
59. 7 bu. 6 qt.
60. 9 gal. 2 qt. 1 pt.
61. 54 years.
62. 94 wk. 6 da.
63. 74 T. 500 lb.
64. 77 yards.
65. 65 mi. 160 rd.
66. 1 da. 14 hr. 18 min.

Page 396.
67. 1323 gal.
68. 4 bu. 2 pk. 4 qt.
69. 2 hr. 34 min. 5 sec.
70. 94 wk. 3 da. 20 hr.
71. 20 bu. 2 pk. 7 qt.
72. 73 yd. 2 ft. 3 in.
73. 21 days.
74. 54 yr. 10 mo.
75. 41 gal. 1 qt.
76. 5 lb. 3 oz.
77. 7 ounces.

78. 3 quarts.
79. 1 bu. 1 pk.
80. 1 hr. 10 min.
81. 5 lb. 13 oz.
82. 18 bu. 3 pk.
7 qt.
83. 16 yd. 2 ft.
9 in.
84. 11 da. 5 hr.
19 min.
85. 93 gal. 3½ qt.
86. 5 hr. 35 min. 5 sec.
87. 22 T. 825 lb.
88. 2 wk. 4 da. 4 hr. 48 min.
89. 18 mi. 180 rd.
90. 5 yr. 9 mo.
91. 13 bu. 3 pk. 6 qt.
92. 25 gal. 2 qt. 1 pt.
93. 33 min. 33 sec.
94. 2 wk. 5 da. 12 hr.
95. 5 yd. 6 in.
96. 7 bu. 2 pk. 1 qt.
97. 1 da. 6 hr. 49 min.
98. 3 qt. 1 pt.
99. 2 yd. 2 ft. 2 in.
100. 10 wk. 2 da. 18 hr. 15 min.
101. 50 hr. 50 min. 50 sec.

102. 11 gal. 1 qt. 1 pt.
103. 22 bu. 2 pk. 2 qt.
104. 17 yd. 1 ft. 9 in.
105. 31 mi. 108 rd. 4 yd.
106. 25 da. 23 hr. 48 min.

Page 397.
108. 13 gal. 1 pt.
110. 17 lb. 3 oz.
111. 4 T. 960 lb.
112. 1 mi. 110 rd.
113. 3 yr. 6 mo.
114. 12 bu. 3 pk. 2⅓ qt.
115. 17 yd. 4 in.
116. 5 hr. 20 min. 10 sec.
117. 18 gal. 1 qt. 1 pt.
118. 3 da. 5 hr. 20 min.
119. 14 T. 110 lb.
120. 16 yd. 2 ft. 11 in.
121. 14 gal. 3 qt. 1½ pt.

Page 399.
1. 1 hr. 18 min. 17 sec.
2. 25 bu. 3 pk. 4 qt.
3. 3½ inches.
4. 2 pk. 6 qt.
5. 14 mi. 17 rd.

ANSWERS.

6. 10 hr. 28 min.
7. 1 ft. 10¼ in.
8. 4 hr. 43 min. 30 sec.
9. 27 min. 10 sec.
10. $37.50.
11. 91¾ cents.
12. 202½ miles.

Page 403.
1. 1129¼.
2. 10,665.
3. 8077 1/11.
4. 28,813⅔.
5. 31,523¼.
6. 61,903 7/11.
7. 206,783¼.
8. 403,270.
9. 834,085½.
10. 15,940,572.
11. 775,665.
12. 933,273.
13. 601,227.
14. 542,817.
15. 2,758,239.
16. 8,296,695.
17. 1,232,766.
18. 3,855,141.
19. 9,733,680.
20. 7,467,570.
21. 67,100.
22. 310,700.
23. 108,662½.
24. 324,133½.
25. 113,437½.
26. 216,500.
27. 426,300.
28. 2150.
29. 105,300.
30. 690,300.

Page 404.
31. 4187 $\frac{4452}{4995}$.
32. 62,132 $\frac{505}{1987}$.
33. 9555 $\frac{2020}{1418}$.
34. 9593 $\frac{2410}{2053}$.
35. 52,073 $\frac{3034}{16074}$.

Page 405.
10. 2 $\frac{5}{14}$.
11. $\frac{2}{15}$.
12. 14⅔.
13. 7¼.
14. 71.01.
15. .89575.
16. 148.28125.
17. .2.
18. $31,370.38.
19. 1 cwt. 3 qr. 10 lb. 10 oz.

---

1. $216,671,399,071.35.

Page 406.
6. $24.90.
7. 21,945 cu. in.
8. $175.
9. $10 gain.
10. $2.47.

Page 407.
4. 20 pounds.
5. 81.
6. $127,581,911,264.12.
7. $1.03 loss.
8. 4800 steps.
9. $120.
10. $912.92.

Page 408.
2. 100 envelopes.
3. 24 rugs.
4. 72 boards.

Page 409.
7. 1944 bricks.
8. 72 tiles.
9. 240 boards.
10. 264,000 stones.
11. 4816 sq. yd.
12. 4840 sq. yd.
13. 80 by 121, 40 × 242, etc.
14. 16 times.
15. 9000 sq. ft.

Page 410.
16. 41,400 sq. ft.; 8600 sq. ft.
17. 9400 sq. ft.
18. $2800; $330.

Page 411.
21. 160 sq. yd.
22. 160 rods.
23. 64 yards.
24. 18 sq. yd.; 20¼ yd.
25. 59 1/16 sq. yd.

---

1. 18 hours.
2. 31 days.

3. 7 o'clock; 3 o'clock; 5 o'clock.
4. $18.

Page 412.
5. 25 cases.
6. 21 posts; 2 posts; 3 posts.
7. 31 days; 29 days.
8. 43 days.
9. 23 chapters.
10. 27 problems.

Page 413.
1. 238 days.
2. 140 days.
3. 109 days.
4. 76 days.
5. 151 days.
6. 284 days.
7. 179 days.
8. 139 days.
9. 91 days.
10. 151 days.
11. $196.
12. 235 days.
13. Tuesday.
14. 66 days.
15. August 15.
16. 170 days.

Page 414.
17. 44 yr. 4 mo. 12 da.
18. 4 yr. 1 mo. 11 da.
19. 8 yr. 4 mo. 14 da.

ANSWERS. 17

20. 128 yr. 2 mo. 9 da.
21. Mar. 4, 1841.
22. 33 yr. 1 mo. 8 da.
23. 3 yr. 9 mo. 15 da.
24. 49 yr. 3 mo. 15 da.
25. July 21, '61.
26. 117 yr. 5 mo. 27 da.

Page 415.
1. $34.56.
2. $15.30.
3. 469 bushels.
4. 108 cows.
5. 216 yards.
6. 192 soldiers.
7. 116 gallons.
8. 31 cents.
9. $39.
10. $1.56.
11. 39 cents.
12. 240 hours.
13. 2 hr. 24 min.
14. $1.77¼.
15. $2.20.
16. 27 days.

Page 416.
1. $34.40.
2. 48 cents.
3. $15.60.
4. $3.90.
5. 35 cents.

Page 417.
6. $10.40.
7. $22.40.

8. $1.
9. $108.
10. $9.60.

Page 418.
1. $\frac{188}{35}$.
2. 11 miles.
3. 3 bu. 7 qt.
4. A, $750; B, $500; C, $250.
5. $39.38.
6. $2.15.
7. $1407.
8. $7.05.
9. 48 tiles.
10. $75.60.
11. $216.66⅔.
12. 12 bu. 3 pk. 4 qt.

Page 419.
1. 18,500 sq. ft.
2. 28 sq. yd.

Page 420.
4. 864 bricks.
5. 1728 bricks.
6. 112 sq. in.
7. 36 sq. ft.
8. 45 rolls.
9. 24 lots.
11. 9000 sq. ft.
12. $1200.

Page 421.
13. 16 fields.
14. 275 yards.
15. 120 sq. rd.
16. 5 acres.
17. 405 sq. yd.

18. 18,755 sq. yd.

Page 422.
19. ½ acre.
20. 160 sq. in.

1. $7\frac{9}{11}$ rods.
2. 7 rd. 4½ yd.
3. 7 rd. 4 yd. 1½ ft.
4. 7 rd. 4 yd. 1 ft. 6 in.
5. 13 rd. 1 ft. 6 in.
6. 12 rods.

Page 423.
7. 8 rd. 5 yd.
8. 8 rd. 5 yd.
9. 8 rd. 5 yd.
10. 8 rd. 5 yd.
11. 9 rd. 1 ft. 6 in.
12. 9 rd. 1 yd. 1 ft. 6 in.
13. 9 rd. 1 yd. 1 ft. 6 in.
14. 9 rd. 1 yd. 1 ft. 6 in.
15. 9 rd. 1 yd. 1 ft. 7 in.
16. 9 rd. 2 yd. 1 ft. 6 in.
17. 9 rods.
18. 9 rods.
19. 9 rods.
20. 9 rods.
21. 7 rd. 2 yd. 2 ft. 1 in.

22. 4 rd. 5 yd. 1 ft.
23. 6 rd. 4 yd. 1 ft. 1 in.
24. 14 rd. 2 ft.
25. 5 rd. 3 yd. 1 ft. 6 in.
26. 7 rd. 2 yd. 1 ft.
27. 17 rd. 2 yd. 1 ft. 3 in.
28. 6 rd. 2 yd. 1 ft. 6 in.
29. 7 rd. 5 yd. 10 in.
30. 990 inches.
31. 5 rods.
32. 1422 inches.
33. 7 rd. 1 yd.

Page 424.
34. 17 rd. 3 yd.
35. 15 rd. 2 yd. 2 ft. 6 in.
36. 13 rods.
37. 22 rd. 3 yd. 2 ft. 6 in.
38. 5 rd. 5 yd. 6 in.
39. 12 rd. 4 yd. 6 in.
40. 23 rd. 2 yd. 6 in.
41. 111 rd. 1 yd. 6 in.
42. 3 rd. 4 yd. 2 ft. 6 in.
43. 3 rd. 4 yd. 1 ft. 6 in.

ANSWERS.

Page 425.
5. 70 cu. in.
8. 46,656 cu. in.
9. $\frac{1}{3}$ cu. yd.
10. 8 ft.×4 ft.;
    16 ft.×2 ft.;
    etc.
11. 5184 cu. ft.
12. 3 × 7 × 11;
    6 × 7 × 5$\frac{1}{2}$;
    etc.
13. 1 cu. ft. smaller.
14. $132.

Page 426.
15. 3 feet.
16. About 7$\frac{1}{2}$ gal.
17. About 1$\frac{1}{4}$ cu. ft.
18. 30 gallons.
19. 30 bushels.
20. 9 cords.
21. 162,000 bricks.
    10,368,000 cu. in.
22. 31,104 bricks.
23. 27 bricks.
24. 40,000 bricks.
25. $2048.

Page 428.
4. 1562.5.
5. $39.81.
8. 88 cents; $\frac{5}{2376}$ rod.
9. $\frac{5}{18}$ bbl.; 127$\frac{7}{8}$ cu. yd.

1. $3.62.

Page 429.
2. $46.55.
3. $7.32.
4. 795 minutes.
5. 1 mi. 85 rd. 2 ft. 6 in.
7. 40 sq. in.
8. .625 year. 643 lb. 9.6 oz.
10. 31.416. 2 inches.

1. 3\frac{5}{12}$.

Page 430.
3. $28,800.
5. $\frac{4}{5}$ ton.
7. $75.47.
9. 40 bushels.
10. $\frac{4}{5}$.

1. $693.
2. 45 cents.
3. 1$\frac{23}{49}$ day.
4. $2.39.
5. 24 miles.

Page 431.
1. $136.57.
2. 60$\frac{20}{21}$ acres.
3. $\frac{375}{3608}$.
4. 5$\frac{14}{15}$.
5. Increased $\frac{1}{15}$.
6. 8.384964.
9. 47 min. 12$\frac{2}{35}$ sec.
10. 1$\frac{49}{88}$.
11. 7.41\frac{3}{4}$.
13. 453$\frac{1}{4}$ miles.
14. 99$\frac{13}{15}$ cents.

Page 432.
15. $16.50.
16. $25.
17. 149$\frac{47}{7}$ gal.
18. $2.51.
19. 2500.
21. $6.66.
24. 23$\frac{52}{103}$.
25. $2.80.

Page 433.
1. 1$\frac{3}{8}$; $\frac{233}{500}$.
2. .571$\frac{3}{7}$; .625.
3. 199.925.
4. .012.
5. $27.
6. 36 spoons.

Page 434.
7. $20.
8. $71.28.
9. 4 feet.
10. $120.
11. 76 sq. yd.
12. $24.75.
13. $23.52.
14. 12,960 lb.
15. $14.28.

Page 435.
3. 10$\frac{17}{50}$; 84$\frac{3}{5}$.
4. 41.00679.
5. 2750 sq. yd.
6. 926$\frac{1}{4}$.

Page 437.
1. 447.77\frac{7}{9}$.
2. 493.76\frac{1}{4}$.
3. $24.75.
4. $141.95.
5. 100.

6. $10.55. 320 feet.
7. 9.37\frac{1}{2}$.
8. 141.
10. $4.50. 5$\frac{55}{141}$ tons.
11. $\frac{8}{11}$ acre.

Page 438.
12. $6; $18; $24; $36.
13. 672 hens.
14. 51$\frac{23}{40}$ miles.
15. $114.60.
16. $1250.
17. 106,294.4. 3757.2.
18. 54 sq. yd. 160 sq. in.
19. $400.
20. 1232 mi.; 1730 yd.; 1,020,304 lb.
21. 24.37\frac{1}{2}$.
22. 173,218.35; 814.43.

Page 439.
23. 3300 ft.; $\frac{27}{45}$ day; 110,672 oz.
24. 31$\frac{11}{24}$ sq. yd.; 42 feet.
25. $\frac{3}{5}$.
26. $714.
27. $9120; $14,820.
28. $\frac{1}{105}$; 278$\frac{4}{40}$.
29. $166.72.

ANSWERS.

30. $60.75.

1. 38 11/12.
2. 37 1/10.
3. 58 11/16.
4. 31 7/8.
5. 61 1/21.
6. 68 7/15.
7. 66 7/15.
8. 95 11/12.
9. 38 11/12.
10. 89 7/10.

Page 440.
11. 6 2/3.
12. 13 11/12.
13. 54 1/2.
14. 67 7/8.
15. 17 1/8.
16. 61 2/3.
17. 37 3/4.
18. 18 7/8.
19. 39 1/2.
20. 25 1/2.
21. 131 1/4.
22. 211 1/4.
23. 663 2/3.
24. 185 5/7.
25. 103 11/12.
26. 95 2/3.
27. 81 1/4.
28. 98 2/3.
29. 513 1/2.
30. 431 3/7.
31. 15 3/4.
32. 13 1/2.
33. 16 11/12.
34. 12 11/12.
35. 21 5/28.
36. 31 44/117.
37. 23 11/117.
38. 22 11/12.
39. 30 41/118.
40. 20 1/15.

Page 441.
1. 460.12.
   21,355.74.
2. .4551.
3. 523/2520.
4. 7 2/3.
5. 11 701/1310.
6. 725 11/12.
7. 27/128.
8. .675.
9. $227.60 5/12.
10. $3800.
11. 23 hr. 2 min. 8 2/3 sec.
12. $11.08 1/4.
13. 1 1/4 yards.
14. 1.
15. $187.50.
16. 7 days.

Page 442.
17. 216 sq. in.;
    1 1/2 sq. ft.;
    216 cu. in.;
    1/8 cu. ft.
18. 28 1/2 feet.
19. 10 2/3 years.
20. $21.67 1/4.
21. $199.50.
22. 7.92 inches.
23. 40 cents.
24. $1800;
    $6300.
25. $874.80.
26. 4142 1/4 lb.
27. 62 1/2 %.

28. $518.40.
29. $183.

Page 445.
1. 150 sq. in.
2. 1536 sq. yd.
3. 117 sq. yd.
4. 1225 sq. ft.
5. 1554 sq. yd.
6. 6111 sq. ft.
7. 92 1/4 sq. m.
8. 81 sq. ft.

Page 446.
9. 81 sq. ft.
10. 735 sq. yd.

Page 448.
2. 20 and 80.
3. $2000;
   $4000;
   $12,000.
4. 18 girls; 36 boys.
5. 13 and 65.
6. 13.

Page 449.
7. 11.
8. $3000;
   $6000;
   $18,000.
9. 12 and 60.
10. 9 marbles;
    18 marbles;
    27 marbles.
11. 36 years;
    6 years.
12. 8.
13. 1; 4; 12; 24.

14. 30; 15; 135.
15. 9 pounds.
16. 19 rods.
17. 85 feet.

Page 450.
18. Son, $40;
    daughter, $80.
19. 25 days.
20. Girl, $80;
    boy, $40.
21. Father, 30 da.;
    son, 15 da.
22. 3 dimes, 6 nickels, 18 cents.
23. 25 yards.
24. 25 rods; 100 rods.
25. Speller, 15 ¢;
    reader, 45 ¢.
26. 60 and 12.
27. 18 nuts; 9 nuts; 27 nuts.

Page 452.
1. 24.
2. 24.
3. 42.
4. 84.
5. 24.
6. 70.
7. 72.
8. 40.
9. 360.
10. 160.
11. 18.
12. 18.
13. 8.
14. 16.

ANSWERS.

15. 12.
16. 20.
17. 900.
18. 60.
19. 60.
20. 32.

Page 453.
21. 36.
22. 222.
23. 180.
24. 72.
25. 320.
26. 7.

1. 15 and 75.
2. 28$\frac{4}{7}$; 71$\frac{3}{7}$.
3. $816.
4. $180.
5. 89.
6. 100.
7. 40; 15.
8. $\frac{3\frac{1}{2}}{2}$.
9. $\frac{4\frac{0}{0}}{4}$.

Page 454.
10. 60; 420.
11. 540; 18.
12. 9.
13. 20 peaches; 5 plums.
14. $200; $600; $700.
15. $60; $140.
16. $300.
17. 64 marbles.
18. $2; $3; $10.
19. $4; $2.
20. 3 horses; 12 cows.

Page 455.
1. 19.
2. 22.
3. 47.
4. 14.
5. 9.
6. 10.
7. 6.
8. 33.

Page 456.
9. 27.
10. 3.
11. 28.
12. 96.
13. 144.
14. 18.
15. 24.
16. 6.
17. 32.
18. 18.
19. 12.
20. 20.

2. 15.
3. 9.
4. 15 marbles; 33 marbles.

Page 457.
5. 25 ft.; 100 ft.
6. 39 acres; 47 acres.
7. 1059 votes; 1377 votes.

8. 62 years.
9. 84; 12.
10. $108.
11. 17; 28.
12. $16; $11.
13. Cows, $45; horses, $125.
14. 3 dimes; 14 half dimes.
15. 74 and 26.
16. 21 boys; 33 girls.

Page 458.
17. $3600; $6000; $8400.
18. 44; 11.
19. 5 five-cent stamps; 20 two-cent stamps; 35 postal cards.
20. 8 horses; 25 cows; 55 sheep.

HISTORY.

Sheldon's General History. For high school and college. The only history following the "seminary" or laboratory plan, now advocated by all leading teachers. Price, $1.60.

Sheldon's Greek and Roman History. Contains the first 250 pages of the above book. Price, $1.00.

Teacher's Manual to Sheldon's History. Puts into the instructor's hand the key to the above system. Price, 80 cents.

Sheldon's Aids to the Teaching of General History. Gives list of essential books for reference library. Price, 10 cents.

Bridgman's Ten Years of Massachusetts. Pictures the development of the Commonwealth as seen in its laws. Price, 75 cents.

Shumway's A Day in Ancient Rome. With 59 illustrations. Should find a place as a *supplementary reader* in every high school class studying Cicero, Horace Tacitus, etc. Price, 75 cents.

Old South Leaflets on U. S. History. Reproductions of important political and historical papers, accompanied by useful notes. Price, 5 cents each. Per hundred, $3.00.

This general series of Old South Leaflets now includes the following subjects: The Constitution of the United States, The Articles of Confederation, The Declaration of Independence, Washington's Farewell Address, Magna Charta, Vane's "Healing Question," Charter of Massachusetts Bay, 1629, Fundamental Orders of Connecticut, 1638, Franklin's Plan of Union, 1754, Washington's Inaugurals, Lincoln's Inaugurals and Emancipation Proclamation, The Federalist, Nos. 1 and 2, The Ordinance of 1787, The Constitution of Ohio, Washington's Letter to Benjamin Harrison, Washington's Circular Letter to the Governors. (38 Leaflets now ready.)

Allen's History Topics. Covers Ancient, Modern, and American history, and gives an excellent list of books of reference. Price, 25 cents.

Fisher's Select Bibliog. of Ecclesiastical History. An annotated list of the most essential books for a Theological student's library. Price, 15 cents.

Hall's Methods of Teaching History. "Its excellence and helpfulness ought to secure it many readers." — *The Nation*. Price, $1.50.

Wilson's The State. Elements of Historical and Practical Politics. A text-book for advanced classes in high schools and colleges on the organization and functions of governments. Retail price, $2.00.

D. C. HEATH & CO., Publishers,
BOSTON, NEW YORK AND CHICAGO.

ENGLISH LANGUAGE.

Hyde's Lessons in English, Book I. For the lower grades. Contains exercises for reproduction, picture lessons, letter writing, *uses* of parts of speech, etc. 40 cts.

Hyde's Lessons in English, Book II. For Grammar schools. Has enough technical grammar for correct use of language. 60 cts.

Hyde's Lessons in English, Book II with Supplement. Has, in addition to the above, 118 pages of technical grammar. 70 cts. Supplement bound alone, 35 cts.

Hyde's Advanced Lessons in English. For advanced classes in grammar schools and high schools. 60 cts.

Hyde's Lessons in English, Book II with Advanced Lessons. The Advanced Lessons and Book II bound together. 80 cts.

Hyde's Derivation of Words. 15 cts.

Mathews's Outline of English Grammar, with Selections for Practice. The application of principles is made through composition of original sentences. 80 cts.

Buckbee's Primary Word Book. Embraces thorough drills in articulation and in the primary difficulties of spelling and sound. 30 cts.

Sever's Progressive Speller. For use in advanced primary, intermediate, and grammar grades. Gives spelling, pronunciation, definition, and use of words. 30 cts.

Badlam's Suggestive Lessons in Language. Being Part I and Appendix of Suggestive Lessons in Language and Reading. 50 cts.

Smith's Studies in Nature, and Language Lessons. A combination of object lessons with language work. 50 cts. Part I bound separately, 25 cts.

Meiklejohn's English Language. Treats salient features with a master's skill and with the utmost clearness and simplicity. $1.30.

Meiklejohn's English Grammar. Also composition, versification, paraphrasing, etc. For high schools and colleges. 90 cts.

Meiklejohn's History of the English Language. 78 pages. Part III of English Language above, 35 cts.

Williams's Composition and Rhetoric by Practice. For high school and college. Combines the smallest amount of theory with an abundance of practice. Revised edition. $1.00.

Strang's Exercises in English. Examples in Syntax, Accidence, and Style for criticism and correction. 50 cts.

Huffcutt's English in the Preparatory School. Presents as practically as possible some of the advanced methods of teaching English grammar and composition in the secondary schools. 25 cts.

Woodward's Study of English. Discusses English teaching from primary school to high collegiate work. 25 cts.

Genung's Study of Rhetoric. Shows the most practical discipline of students for the making of literature. 25 cts.

Goodchild's Book of Stops. Punctuation in Verse. Illustrated. 10 cts.

See also our list of books for the study of English Literature.

D. C. HEATH & CO., PUBLISHERS,
BOSTON. NEW YORK. CHICAGO.

EDUCATION.

Compayré's History of Pedagogy. "The best and most comprehensive history of Education in English." — Dr. G. S. HALL. $1.75.

Compayré's Lectures on Teaching. "The best book in existence on the theory and practice of education." — Supt. MACALISTER, Philadelphia. $1.75.

Compayré's Psychology Applied to Education. A clear and concise statement of doctrine and application on the science and art of teaching. 90 cts.

De Garmo's Essentials of Method. A practical exposition of methods with illustrative outlines of common school studies. 65 cts.

De Garmo's Lindner's Psychology. The best Manual ever prepared from the Herbartian standpoint. $1.00.

Gill's Systems of Education. "It treats ably of the Lancaster and Bell movement in education, — a *very important* phase." — Dr. W. T. HARRIS. $1.25.

Hall's Bibliography of Pedagogical Literature. Covers every department of education. Interleaved, *$2.00. $1.50.

Herford's Student's Froebel. The purpose of this little book is to give young people preparing to teach a brief yet full account of Froebel's Theory of Education. 75 cts.

Malleson's Early Training of Children. "The best book for mothers I ever read." — ELIZABETH P. PEABODY. 75 cts.

Marwedel's Conscious Motherhood. The unfolding of the child's mind in the cradle, nursery and Kindergarten. $2.00.

Newsholme's School Hygiene. Already in use in the leading training colleges in England. 75 cts.

Peabody's Home, Kindergarten, and Primary School. "The best book outside of the Bible that I ever read." — A LEADING TEACHER. $1.00.

Pestalozzi's Leonard and Gertrude. "If we except 'Emile' only, no more important educational book has appeared for a century and a half than 'Leonard and Gertrude.'" — *The Nation.* 90 cts.

Radestock's Habit in Education. "It will prove a rare 'find' to teachers who are seeking to ground themselves in the philosophy of their art." — E. H. RUSSELL, Worcester Normal School. 75 cts.

Richter's Levana; or, The Doctrine of Education. "A spirited and scholarly book." — Prof. W. H. PAYNE. $1.40.

Rosmini's Method in Education. "The most important pedagogical work ever written." — THOMAS DAVIDSON. $1.50.

Rousseau's Emile. "Perhaps the most influential book ever written on the subject of Education." — R. H. QUICK. 90 cts.

Methods of Teaching Modern Languages. Papers on the value and on methods of teaching German and French, by prominent instructors. 90 cts.

Sanford's Laboratory Course in Physiological Psychology. The course includes experiments upon the Dermal Senses, Static and Kinæsthetic Senses, Taste, Smell, Hearing, Vision, Psychophysic. *In Press.*

Lange's Apperception: A monograph on Psychology and Pedagogy. Translated by the members of the Herbart Club, under the direction of President Charles DeGarmo, of Swarthmore College. $1.00.

Herbart's Science of Education. Translated by Mr. and Mrs. Felken with a preface by Oscar Browning. $1.00.

Tracy's Psychology of Childhood. This is the first *general* treatise covering in a scientific manner the whole field of child psychology. Octavo. Paper. 75 cts.

Sent by mail, postpaid, on receipt of price.

D. C. HEATH & CO., PUBLISHERS,
BOSTON. NEW YORK. CHICAGO.

READING.

Badlam's Suggestive Lessons in Language and Reading. A manual for primary teachers. Plain and practical; being a transcript of work actually done in the school-room. $1.50.

Badlam's Stepping-Stones to Reading.— A Primer. Supplements the 283-page book above. Boards. 30 cts.

Badlam's First Reader. New and valuable word-building exercises, designed to follow the above. Boards. 35 cts.

Bass's Nature Stories for Young Readers: Plant Life. Intended to supplement the first and second reading-books. Boards. 30 cts.

Bass's Nature Stories for Young Readers: Animal Life. Gives lessons on animals and their habits. To follow second reader. Boards. 40 cts.

Fuller's Illustrated Primer. Presents the word-method in a very attractive form to the youngest readers. Boards. 30 cts.

Fuller's Charts. Three charts for exercises in the elementary sounds, and for combining them to form syllables and words. The set for $1.25. Mounted, $2.25.

Hall's How to Teach Reading. Treats the important question: what children should and should not read. Paper. 25 cts.

Miller's My Saturday Bird Class. Designed for use as a supplementary reader in lower grades or as a text-book of elementary ornithology. Boards. 30 cts.

Norton's Heart of Oak Books. This series is of material from the standard imaginative literature of the English language. It draws freely upon the treasury of favorite stories, poems, and songs with which every child should become familiar, and which have done most to stimulate the fancy and direct the sentiment of the best men and women of the English-speaking race. Book I, 100 pages, 25 cts.; Book II, 142 pages, 35 cts.; Book III, 265 pages, 45 cts.; Book IV, 303 pages, 55 cts.; Book V, 359 pages, 65 cts.; Book VI, 367 pages, 75 cts.

Smith's Reading and Speaking. Familiar Talks to those who would speak well in public. 70 cts.

Spear's Leaves and Flowers. Designed for supplementary reading in lower grades or as a text-book of elementary botany. Boards. 30 cts.

Ventura's Mantegazza's Testa. A book to help boys toward a complete self-development. $1.00.

Wright's Nature Reader, No. I. Describes crabs, wasps, spiders, bees, and some univalve mollusks. Boards. 30 cts.

Wright's Nature Reader, No. II. Describes ants, flies, earth-worms, beetles, barnacles and star-fish. Boards. 40 cts.

Wright's Nature Reader, No. III. Has lessons in plant-life, grasshoppers, butterflies, and birds. Boards. 60 cts.

Wright's Nature Reader, No. IV. Has lessons in geology, astronomy, world-life, etc. Boards. 70 cts.

For advanced supplementary reading see our list of books in English Literature.

D. C. HEATH & CO., PUBLISHERS,
BOSTON.   NEW YORK.   CHICAGO.

NUMBER.

White's Two Years with Numbers. Number Lessons for second and third year pupils. 40 cts.

Atwood's Complete Graded Arithmetic. Present a carefully graded course in arithmetic, to begin with the fourth year and continue through the eighth year. Part I. 200 pages. Cloth. 40 cts. Part II. 382 pages. Half leather. 75 cts.

Walsh's Mathematics for Common Schools. Special features of this work are its division into half-yearly chapters instead of the arrangement by topics; the omission, as far as possible, of rules and definitions; the great number and variety of the problems; the use of the equation in solution of arithmetical problems; and the introduction of the elements of algebra and geometry. Part I. 218 pages. 35 cts. Part. II. 252 pages. 40 cts. Part III. 365 pages. Half leather. 75 cts.

Sutton and Kimbrough's Pupils' Series of Arithmethics.
PRIMARY BOOK. Embraces the four fundamental operations in all their simple relations. 80 pages. Boards. 22 cts.
INTERMEDIATE BOOK. Embraces practical work through the four operations cancellation, factoring and properties of numbers, simple and decimal fractions, percentage and simple interest. 128 pages. Boards. 25 cts.
LOWER BOOK. Combines in one volume the Primary and Intermediate Books. 208 pages. Boards, 30 cts. Cloth, 45 cts.
HIGHER BOOK. A compact volume for efficient work which makes clear all necessary theory. 275 pages. Half leather. 70 cts.

Safford's Mathematical Teaching. Presents the best methods of teaching, from primary arithmetic to the calculus. Paper. 25 cts.

Badlam's Aids to Number. *For Teachers. First Series.* Consists of 25 cards for sight-work with objects from one to ten. 40 cts.

Badlam's Aids to Number. *For Pupils. First Series.* Supplements the above with material for slate work. Leatherette. 30 cts.

Badlam's Aids to Number. *For Teachers. Second Series.* Teachers' sight-work with objects above ten. 40 cts.

Badlam's Aids to Number. *For Pupils. Second Series.* Supplements above with material for slate work from 10 to 20. Leatherette. 30 cts.

Badlam's Number Chart. 11 x 14 inches. Designed to aid in teaching the four fundamental rules in lowest primary grades. 5 cts. each; per hundred $4.00.

Luddington's Picture Problems. 70 cards, 3 x 5 inches, in colors, to teach by pictures combinations from one to ten. 65 cts.

Pierce's Review Number Cards. Two cards, 7 x 9, for rapid work for second and third year pupils. 3 cts. each; per hundred $2.40.

Howland's Drill Card. For rapid practice work in middle grades. 3 cts. each; per hundred $2.40.

For advanced work see our list of books in Mathematics.

D. C. HEATH & CO., PUBLISHERS,
BOSTON.   NEW YORK.   CHICAGO.

ALGEBRA AND GEOMETRY.

Academic Algebra.

By E. A. BOWSER, Prof. of Mathematics, Rutgers College. Half leather. 366 pages. Price by mail, $1.25.

This work is designed as a text-book for common and high schools and academies, and to prepare students for entering colleges and scientific schools. The book is a complete treatise on Algebra up to and through the Progressions, and including Permutations and Combinations and the Binomial Theorem. For students who have not sufficient time to take the College Algebra, this perhaps is the better book; but those who contemplate entering college, or who wish to take a complete course in Algebra, may as well begin at once with the larger work.

College Algebra.

Half leather. 558 pages. Price by mail, $1.65. Introduction price, $1.50.

This work is designed as a text-book for academies, colleges, and scientific schools. It begins at the beginning of the subject, and the full treatment of the earlier parts renders it unnecessary that students who use it shall have previously studied a more elementary algebra.

Plane and Solid Geometry.

Half leather. 402 pages. Price by mail, $1.40. Introduction price, $1.25.

This work combines the excellences of Euclid with those of the best modern writers, especially of Legendre and Rouché and De Comberousse. It aims to effect two objects: (1) to teach geometric truths; (2) to discipline and invigorate the mind — to train it to habits of clear and consecutive reasoning.

Hopkins' Plane Geometry, on the Heuristic Plan.

By G. I. HOPKINS, High School, Manchester, N.H. Boards. 60 cents.

The demonstrations are purposely incomplete, so that the pupil is compelled to master the subject instead of memorizing it.

D. C. HEATH & CO., Publishers.

BOSTON.   NEW YORK.   CHICAGO.

ARITHMETIC.

Aids to Number. — First Series. Teachers' Edition.

Oral Work — One to ten. 25 cards with concise directions. By ANNA B. BADLAM, Principal of Training School, Lewiston, Me., formerly of Rice Training School, Boston. Retail price, 40 cents.

Aids to Number. — First Series. Pupils' Edition.

Written work. — One to ten. Leatherette. Introduction price, 25 cents.

Aids to Number. — Second Series. Teachers' Edition.

Oral Work. — Ten to One Hundred. With especial reference to multiples of numbers from 1 to 10. 32 cards with concise directions. Retail price, 40 cents.

Aids to Numbers. — Second Series. Pupils' Edition.

Written Work. — Ten to One Hundred. Leatherette. Introduction price, 25 cents.

The Child's Number Charts. By ANNA B. BADLAM.

Manilla card, 11 x 14 inches. Price, 5 cents each; $4.00 per hundred.

Drill Charts. By C. P. HOWLAND, Principal of Tabor Academy, Marion, Mass.

For rapid, middle-grade practice work on the Fundamental Rules of Arithmetic. Two cards, 8 x 9 inches. Price, 3 cents each; or $2.40 per hundred.

Review Number Cards. By ELLA M. PIERCE, of Providence, R. I.

For Second and Third Year Pupils. Cards, 7 x 9 inches. Price, 3 cents each; or $2.40 per hundred.

Picture Problems. By MISS H. A. LUDDINGTON,

Principal of Training School, Pawtucket, R. I.; formerly Teacher of Methods and Training Teacher in Primary Department of State Normal School, New Britain, Conn., and Training Teacher in Cook County Normal School, Normal Park, Ill. 70 colored cards, 4 x 5 inches, printed on both sides, arranged in 9 sets, 6 to 10 cards in each set, with card of directions. Retail price, 65 cents.

Mathematical Teaching and its Modern Methods.

By TRUMAN HENRY SAFFORD, Ph. D., Professor of Astronomy, Williams College, Mass. Paper. 47 pages. Retail price, 25 cents.

The New Arithmetic.

By 300 authors. Edited by SEYMOUR EATON, with Preface by T. H. SAFFORD, Professor of Astronomy, Williams College, Mass. Introduction price, 75 cents.

D. C. HEATH & CO., Publishers,
BOSTON, NEW YORK, AND CHICAGO.

MATHEMATICS.

Bowser's Academic Algebra. A complete treatise through the progressions, including Permutations, Combinations, and the Binomial Theorem. Half leather. $1.25.

Bowser's College Algebra. A complete treatise for colleges and scientific schools. Half leather. $1.65.

Bowser's Plane and Solid Geometry. Combines the excellences of Euclid with those of the best modern writers. Half leather. $1.35.

Bowser's Plane Geometry. Half leather. 85 cts.

Bowser's Elements of Plane and Spherical Trigonometry. A brief course prepared especially for High Schools and Academies. Half leather. $1.00.

Bowser's Treatise on Plane and Spherical Trigonometry. An advanced work which covers the entire course in higher institutions. Half leather. $1.65.

Hanus's Geometry in the Grammar Schools. An essay, together with illustrative class exercises and an outline of the work for the last three years of the grammar school. 52 pages. 25 cts.

Hopkin's Plane Geometry. On the heuristic plan. Half leather. 85 cts.

Hunt's Concrete Geometry for Grammar Schools. The definitions and elementary concepts are to be taught concretely, by much measuring, by the making of models and diagrams by the pupil, as suggested by the text or by his own invention. 100 pages. Boards. 30 cts.

Waldo's Descriptive Geometry. A large number of problems systematically arranged and with suggestions. 90 cts.

The New Arithmetic. By 300 teachers. Little theory and much practice. Also an excellent review book. 230 pages. 75 cts.

For Arithmetics and other elementary work see our list of books in Number.

D. C. HEATH & CO., PUBLISHERS,
BOSTON.   NEW YORK.   CHICAGO.

DRAWING AND MANUAL TRAINING.

Johnson's Progressive Lessons in Needlework. Explains needlework from its rudiments and gives with illustrations full directions for work during six grades. 117 pages. Square 8vo. Cloth, $1.00. Boards, 60 cts.

Seidel's Industrial Instruction (Smith). A refutation of all objections raised against industrial instruction. 170 pages. 90 cts.

Thompson's Educational and Industrial Drawing.
    Primary Free-Hand Series (Nos. 1-4). Each No., per doz., $1.00.
    Primary Free-Hand Manual. 114 pages. Paper. 40 cts.
    Advanced Free-Hand Series (Nos. 5-8). Each No., per doz., $1.50.
    Model and Object Series (Nos. 1-3). Each No., per doz., $1.75.
    Model and Object Manual. 84 pages. Paper. 35 cts.
    Æsthetic Series (Nos. 1-6). Each No., per doz., $1.50.
    Æsthetic Manual. 174 pages. Paper. 60 cts.
    Mechanical Series (Nos. 1-6). Each No., per doz., $2.00.
    Mechanical Manual. 172 pages. Paper. 75 cts.
    Models to accompany Thompson's Drawing:
    Set No. I. For Primary Books, per set, 40 cts.
    Set No. II. For Model and Object Book No. 1, per set, 00 cts.
    Set No. III. For Model and Object Book No. 2, per set, 50 cts.

Thompson's Manual Training, No. 1. Treats of Clay Modelling, Stick and Tablet Laying, Paper Folding and Cutting, Color, and Construction of Geometrical Solids. Illustrated. 66 pages. Large 8vo. Paper. 30 cts.

Thompson's Manual Training, No. 2. Treats of Mechanical Drawing, Clay-Modelling in Relief, Color, Wood Carving, Paper Cutting and Pasting. Illustrated. 70 pp. Large 8vo. Paper. 30 cts.

Waldo's Descriptive Geometry. A large number of problems systematically arranged, with suggestions. 85 pages. 90 cts.

Whitaker's How to Use Wood Working Tools. Lessons in the uses of the universal tools: the hammer, knife, plane, rule, chalk-line, square, gauge, chisel, saw, and auger. 104 pages. 60 cts.

Woodward's Manual Training School. Its aims, methods, and results; with detailed courses of instruction in shop-work. Fully illustrated. 374 pages. Octavo. $2.00.

Woodward's Educational Value of Manual Training. Sets forth more clearly and fully than has ever been done before the true character and functions of manual training in education. 96 pages. Paper. 25 cts.

Sent postpaid by mail on receipt of price.

D. C. HEATH & CO., PUBLISHERS,
BOSTON. NEW YORK. CHICAGO.

Why should Teachers Read the Literature of their Profession?

1. Because *no man can stand high in any profession who is not familiar with its history and literature.*

2. Because *it saves time which might be wasted in trying experiments that have already been tried and found useless.*

Compayré's History of Pedagogy. "The best and most comprehensive history of Education in English." — Dr. G. S. HALL.	$1.75
Compayré's Lectures on Teaching. "The best book in existence on the theory and practice of Education." — Supt. MACALLISTER, Philadelphia.	1.75
Gill's System of Education. "It treats ably of the Lancaster and Bell movement in Education — a very important phase." — Dr. W. T. HARRIS.	1.25
Radestock's Habit in Education. "It will prove a rare 'find' to teachers who are seeking to ground themselves in the philosophy of their art." — E. H. RUSSELL, Worcester Normal.	0.75
Rousseau's Emile. "Perhaps the most influential book ever written on the subject of Education." — R. H. QUICK.	0.90
Pestalozzi's Leonard and Gertrude. "If we except 'Emile' only, no more important educational book has appeared, for a century and a half, than 'Leonard and Gertrude.'" — *The Nation*.	0.90
Richter's Levana; or the Doctrine of Education. "A spirited and scholarly book." — Prof. W. H. PAYNE.	1.40
Rosmini's Method in Education. "The most important pedagogical work ever written." — THOMAS DAVIDSON.	1.50
Malleson's Early Training of Children. "The best book for mothers I ever read." — ELIZABETH P. PEABODY.	0.75
Hall's Bibliography of Pedagogical Literature. Covers every department of Education.	1.50
Peabody's Home, Kindergarten and Primary School Education. "The best book outside of the Bible I ever read." — A LEADING TEACHER.	1.00
Newsholme's School Hygiene. Already in use in the leading training colleges in England.	0.75
DeGarmo's Essentials of Method. "It has as much sound thought to the square inch as anything I know of in pedagogics." — Supt. BALLIET, Springfield, Mass.	0.65
Hall's Methods of Teaching History. "Its excellence and helpfulness ought to secure it many readers." — *The Nation*.	1.50
Seidel's Industrial Education. "It answers triumphantly all objections to the introduction of manual training to the public schools." — CHARLES H. HAM, Chicago.	0.90
Badlam's Suggestive Lessons on Language and Reading. "The book is all that it claims to be and more. It abounds in material that will be of service to the progressive teacher." — Supt. DUTTON, New Haven.	1.50
Redway's Teachers' Manual of Geography. "Its hints to teachers are invaluable, while its chapters on 'Modern Facts and Ancient Fancies' will be a revelation to many." — ALEX. E. FRYE, Author of "*The Child in Nature*."	0.65
Nichols' Topics in Geography. "Contains excellent hints and suggestions of incalculable aid to school teachers." — *Oakland (Cal.) Tribune*.	0.65

D. C. HEATH & CO., Publishers,
BOSTON, NEW YORK AND CHICAGO.

BUSINESS.

Seavy's Practical Business Bookkeeping. All needless discussion is carefully avoided. Only such explanations are given as are essential to preparation for actual business duties. Half leather. $1.65.

Blanks to Accompany Seavy's Practical Business Bookkeeping. Per set of three, 70 cts.

Seavy's Manual of Business Transactions. Contains transactions for practice, together with instructions and references to the author's Bookkeeping. 45 cts.

Shaw's Practice Book of Business Forms and Elementary Bookkeeping. Treats of the best methods of keeping simple accounts and furnishes a necessary knowledge of ordinary business forms. Flexible boards. 70 cts.

Blanks to Accompany Shaw's Practice Book of Business Forms.
    Boards 24
    Blanks for single entry. Per set of three 30
    Book of Blank Notes, Bill Heads, Bank Checks, Receipts, Orders, etc. . . .20

Weed's Business Law. A brief statement of the laws that govern business. $1.10.

Heath's Writing Books. (Haaren and Stebbins.) *In press.*

The Volpenna Vertical Writing Books. (Newlands and Row). *In press.*

The New Arithmetic. An excellent review and practice book. 230 pages. 75 cts

---

D. C. HEATH & CO., PUBLISHERS,
BOSTON. NEW YORK. CHICAGO.

GEOGRAPHY AND MAPS.

Heath's Practical School Maps. Each 30 x 40 inches. Printed from new plates and showing latest political changes. The common school set consists of Hemispheres, No. America, So. America, Europe, Africa, Asia, United States. Eyeletted for hanging on wall, singly, $1.25; per set of seven, $7.00. Mounted on cloth and rollers. Singly, $2.00. Mounted on cloth per set of seven, $12.00. Sunday School set. Canaan and Palestine. Singly, $1.25; per set of two, $2.00. Mounted, $2.00 each.

Heath's Outline Map of the United States. Invaluable for marking territorial growth and for the graphic representation of all geographical and historical matter. Small (desk) size, 2 cents each; $1.50 per hundred. Intermediate size, 30 cents each. Large size, 50 cts.

Historical Outline Map of Europe. 12 x 18 inches, on bond paper, in black outline. 3 cents each; per hundred, $2.25.

Jackson's Astronomical Geography. Simple enough for grammar schools. Used for a brief course in high school. 40 cts.

Map of Ancient History. Outline for recording historical growth and statistics (14 x 17 in.), 3 cents each; per 100, $2.25.

Nichols' Topics in Geography. A guide for pupils' use from the primary through the eighth grade. 65 cts.

Picturesque Geography. 12 lithograph plates, 15 x 20 inches, and pamphlet describing their use. Per set, $3.00; mounted, $5.00.

Progressive Outline Maps: United States, *World on Mercator's Projection (12 x 20 in.); North America, South America, Europe, *Central and Western Europe, Africa, Asia, Australia, *British Isles, *England, *Greece, *Italy, New England, Middle Atlantic States, Southern States, Southern States—western section, Central Eastern States, Central Western States, Pacific States, New York, Ohio, The Great Lakes, Washington (State), *Palestine (each 10 x 12 in.). For the graphic representation by the pupil of geography, geology, history, meteorology, economics, and statistics of all kinds. 2 cents each; per hundred, $1.50.

Those marked with Star (*) are also printed in black outline for use in teaching history.

Redway's Manual of Geography. I. Hints to Teachers; II. Modern Facts and Ancient Fancies. 65 cts.

Redway's Reproduction of Geographical Forms. I. Sand and Clay-Modelling; II. Map Drawing and Projection. Paper. 30 cts.

Roney's Student's Outline Map of England. For use in English History and Literature, to be filled in by pupils. 5 cts.

Trotter's Lessons in the New Geography. Treats geography from the human point of view. Adapted for use as a text-book or as a reader. *In press.*

D. C. HEATH & CO., PUBLISHERS.
BOSTON.   NEW YORK.   CHICAGO.

ELEMENTARY SCIENCE.

Bailey's Grammar School Physics. A series of inductive lessons in the elements of the science. *In press.*

Ballard's The World of Matter. A guide to the study of chemistry and mineralogy; adapted to the general reader, for use as a text-book or as a guide to the teacher in giving object-lessons. 264 pages. Illustrated. $1.00.

Clark's Practical Methods in Microscopy. Gives in detail descriptions of methods that will lead the careful worker to successful results. 233 pages. Illustrated. $1.60.

Clarke's Astronomical Lantern. Intended to familiarize students with the constellations by comparing them with fac-similes on the lantern face. With seventeen slides, giving twenty-two constellations. $4.50.

Clarke's How to find the Stars. Accompanies the above and helps to an acquaintance with the constellations. 47 pages. Paper. 15 cts.

Guides for Science Teaching. Teachers' aids in the instruction of Natural History classes in the lower grades.
  I. Hyatt's About Pebbles. 26 pages. Paper. 10 cts.
  II. Goodale's A Few Common Plants. 61 pages. Paper. 20 cts.
  III. Hyatt's Commercial and other Sponges. Illustrated. 43 pages. Paper. 20 cts.
  IV. Agassiz's First Lessons in Natural History. Illustrated. 64 pages. Paper. 25 cts.
  V. Hyatt's Corals and Echinoderms. Illustrated. 32 pages. Paper. 30 cts.
  VI. Hyatt's Mollusca. Illustrated. 65 pages. Paper. 30 cts.
  VII. Hyatt's Worms and Crustacea. Illustrated. 68 pages. Paper. 30 cts.
  VIII. Hyatt's Insecta. Illustrated. 324 pages. Cloth. $1.25.
  XII. Crosby's Common Minerals and Rocks. Illustrated. 200 pages. Paper, 40 cts. Cloth, 60 cts.
  XIII. Richard's First Lessons in Minerals. 50 pages. Paper. 10 cts.
  XIV. Bowditch's Physiology. 58 pages. Paper. 20 cts.
  XV. Clapp's 36 Observation Lessons in Minerals. 80 pages. Paper. 30 cts.
  XVI. Phenix's Lessons in Chemistry. *In press.*
  Pupils Note-Book to accompany No. 15. 10 cts.

Rice's Science Teaching in the School. With a course of instruction in science for the lower grades. 46 pages. Paper. 25 cts.

Ricks's Natural History Object Lessons. Supplies information on plants and their products, on animals and their uses, and gives specimen lessons. Fully illustrated. 332 pages. $1.50.

Ricks's Object Lessons and How to Give them.
  Volume I. Gives lessons for primary grades. 200 pages. 90 cts.
  Volume II. Gives lessons for grammar and intermediate grades. 212 pages. 90 cts.

Shaler's First Book in Geology. For high school, or highest class in grammar school 272 pages. Illustrated. $1.00.

Shaler's Teacher's Methods in Geology. An aid to the teacher of Geology 74 pages. Paper. 25 cts.

Smith's Studies in Nature. A combination of natural history lessons and language work. 48 pages. Paper. 15 cts.

Sent by mail postpaid on receipt of price. See also our list of books in Science

---

D. C. HEATH & CO., PUBLISHERS,
BOSTON.　NEW YORK.　CHICAGO.

SCIENCE.

Shaler's First Book in Geology. For high school, or highest class in grammar school. $1.10. Bound in boards for supplementary reader. 70 cts.

Ballard's World of Matter. A Guide to Mineralogy and Chemistry. $1.00.

Shepard's Inorganic Chemistry. Descriptive and Qualitative; experimental and inductive; leads the student to observe and think. For high schools and colleges. $1.25.

Shepard's Briefer Course in Chemistry; with Chapter on Organic Chemistry. Designed for schools giving a half year or less to the subject, and schools limited in laboratory facilities. 90 cts.

Shepard's Organic Chemistry. The portion on organic chemistry in Shepard's Briefer Course is bound in paper separately. Paper. 30 cts.

Shepard's Laboratory Note-Book. Blanks for experiments: tables for the reactions of metallic salts. Can be used with any chemistry. Boards. 40 cts.

Benton's Guide to General Chemistry. A manual for the laboratory. 40 cts.

Remsen's Organic Chemistry. An Introduction to the Study of the Compounds of Carbon. For students of the pure science, or its application to arts. $1.30.

Orndorff's Laboratory Manual. Containing directions for a course of experiments in Organic Chemistry, arranged to accompany Remsen's Chemistry. Boards. 40 cts.

Coit's Chemical Arithmetic. With a short system of Elementary Qualitative Analysis. For high schools and colleges. 60 cts.

Grabfield and Burns' Chemical Problems. For preparatory schools. 60 cts.

Chute's Practical Physics. A laboratory book for high schools and colleges studying physics experimentally. Gives free details for laboratory work. $1.25.

Colton's Practical Zoology. Gives a clear idea of the subject as a whole, by the careful study of a few typical animals. 90 cts.

Boyer's Laboratory Manual in Elementary Biology. A guide to the study of animals and plants, and is so constructed as to be of no help to the pupil unless he actually studies the specimens.

Clark's Methods in Microscopy. This book gives in detail descriptions of methods that will lead any careful worker to successful results in microscopic manipulation. $1.60.

Spalding's Introduction to Botany. Practical Exercises in the Study of Plants by the laboratory method. 90 cts.

Whiting's Physical Measurement. Intended for students in Civil, Mechanical and Electrical Engineering, Surveying, Astronomical Work, Chemical Analysis, Physical Investigation, and other branches in which accurate measurements are required.
   I. Fifty measurements in Density, Heat, Light, and Sound. $1.30.
   II. Fifty measurements in Sound, Dynamics, Magnetism, Electricity. $1.30.
   III. Principles and Methods of Physical Measurement, Physical Laws and Principles, and Mathematical and Physical Tables. $1.30.
   IV. Appendix for the use of Teachers, including examples of observation and reduction. Part IV is needed by students only when working without a teacher. $1.30.
   Parts I-III, in one vol., $3.25. Parts I-IV, in one vol., $4.00.

Williams's Modern Petrography. An account of the application of the microscope to the study of geology. Paper. 25 cts.

For elementary works see our list of books in Elementary Science.

D. C. HEATH & CO., PUBLISHERS.
BOSTON.   NEW YORK.   CHICAGO.

ENGLISH LITERATURE.

Hawthorne and Lemmon's American Literature. A manual for high schools and academies. $1.25.

Meiklejohn's History of English Language and Literature. For high school and colleges. A compact and reliable statement of the essentials; also included in Meiklejohn's English Language (see under English Language). 90 cts.

Meiklejohn's History of English Literature. 116 pages. Part IV of English Literature, above. 45 cts.

Hodgkins' Studies in English Literature. Gives full lists of aids for laboratory method. Scott, Lamb, Wordsworth, Coleridge, Byron, Shelley, Keats, Macaulay, Dickens, Thackeray, Robert Browning, Mrs. Browning, Carlyle, George Eliot, Tennyson, Rossetti, Arnold, Ruskin, Irving, Bryant, Hawthorne, Longfellow, Emerson, Whittier, Holmes, and Lowell. A separate pamphlet on each author. Price 5 cts. each, or per hundred, $3.00; complete in cloth (adjustable file cover, $1.50). $1.00.

Scudder's Shelley's Prometheus Unbound. With Introduction and copious notes. 70 cts.

George's Wordsworth's Prelude. Annotated for high school and college. Never before published alone. 80 cts.

George's Selections from Wordsworth. 168 poems chosen with a view to illustrate the growth of the poet's mind and art. $1.00.

George's Wordsworth's Prefaces and Essays on Poetry. Contains the best of Wordsworth's prose. 60 cts.

George's Webster's Speeches. Nine select speeches with notes. $1.50.

George's Burke's American Orations. Cloth. 65 cts.

George's Syllabus of English Literature and History. Shows in parallel columns, the progress of History and Literature. 20 cts.

Corson's Introduction to Browning. A guide to the study of Browning's Poetry. Also has 33 poems with notes. $1.50.

Corson's Introduction to the Study of Shakespeare. A critical study of Shakespeare's art, with examination questions. $1.50.

Corson's Introduction to the Study of Milton. *In press.*

Corson's Introduction to the Study of Chaucer. *In press.*

Cook's Judith. The Old English epic poem, with introduction, translation, glossary and fac-simile page. $1.60. Students' edition without translation. 35 cts.

Cook's The Bible and English Prose Style. Approaches the study of the Bible from the literary side. 60 cts.

Simonds' Sir Thomas Wyatt and his Poems. 168 pages. With biography, and critical analysis of his poems. 75 cts.

Hall's Beowulf. A metrical translation. $1.00. Students' edition. 35 cts.

Norton's Heart of Oak Books. A series of five volumes giving selections from the choicest English literature.

Phillips's History and Literature in Grammar Grades. An essay showing the intimate relation of the two subjects. 15 cts.

See also our list of books for the study of the English Language.

D. C. HEATH & CO., PUBLISHERS.
BOSTON.   NEW YORK.   CHICAGO.

CIVICS, ECONOMICS, AND SOCIOLOGY.

Boutwell's The Constitution of the United States at the End of the First Century. Contains the Organic Laws of the United States, with references to the decisions of the Supreme Court which elucidate the text, and an historical chapter reviewing the steps which led to the adoption of these Organic Laws. *In press.*

Dole's The American Citizen. Designed as a text-book in Civics and morals for the higher grades of the grammar school as well as for the high school and academy. Contains Constitution of United States, with analysis. 336 pages. $1.00.

Special editions are made for Illinois, Indiana, Ohio, Missouri, Nebraska, No. Dakota, So. Dakota, Wisconsin, Minnesota, Kansas, Texas.

Goodale's Questions to Accompany Dole's The American Citizen. Contains, beside questions on the text, suggestive questions and questions for class debate. 87 pages. Paper. 25 cts.

Gide's Principles of Political Economy. Translated from the French by Dr. Jacobsen of London, with introduction by Prof. James Bonar of Oxford. 598 pages. $2.00.

Henderson's Introduction to the Study of Dependent, Defective, and Delinquent Classes. Adapted for use as a text-book, for personal study, for teachers' and ministers' institutes, and for clubs of public-spirited men and women engaged in considering some of the gravest problems of society. 287 pages. $1.50.

Hodgin's Indiana and the Nation. Contains the Civil Government of the State, as well as that of the United States, with questions. 198 pages. 70 cts.

Lawrence's Guide to International Law. A brief outline of the principles and practices of International Law. *In press.*

Wenzel's Comparative View of Governments. Gives in parallel columns comparisons of the governments of the United States, England, France, and Germany. 26 pages. Paper. 22 cts.

Wilson's The State. Elements of Historical and Practical Politics. A text-book on the organization and functions of government for high schools and colleges. 720 pages. $2.00.

Wilson's United States Government. For grammar and high schools. 140 pages. 60 cts.

Woodburn and Hodgin's The American Commonwealth. Contains several orations from Webster and Burke, with analyses, historical and explanatory notes, and studies of the men and periods. 586 pages. $1.50.

Sent by mail, post paid on receipt of prices. See also our list of books in History.

D. C. HEATH & CO., PUBLISHERS,
BOSTON. NEW YORK. CHICAGO.

www.ingramcontent.com/pod-product-compliance
Lightning Source LLC
Chambersburg PA
CBHW020112170426
43199CB00009B/501